# Sheaf Theory

Yong Seung Cho

KYUNG MOON Publishers

# PREFACE

Sheaf theory has had profound effects on several mathematical subjects, especially, topology, differential and algebraic geometry. It gives a tool for dealing with problems by piecing together solutions of local problems in a coherent manner to get global solutions. For example, several geometric structures of a manifold, scheme on the space and general cohomology theory can be expressed in terms of sheaves. The sheaf cohomology measures the lack of exactness of the global solutions.

In this book we introduce the foundations of sheaf theory to define the cohomologies of topological, differential and complex manifolds. In Chapter 1 we introduce Riemann surfaces and motivation of sheaf cohomology. In Chapter 2 we introduce direct limits, and in Chapter 3 presheaf, sheaf and examples. In Chapter 4 we introduce ringed spaces, geometric spaces and module over ringed spaces. In Chapter 5 we deal with Čech cohomology and Grothendieck cohomology. In Chapter 6 applications to Riemann surfaces. In Chapter 7 complex manifolds, in Chapter 8 the de Rham theory, and in Chapter 9 Hodge decomposition theorems on compact oriented manifolds and Kähler manifolds. In Appendix we introduce characteristic classes, low dimensional manifolds, and categories.

The final manuscript and galley proof are read by my students Ahram Lim and Semin Yoo. I am sincerely grateful to them.

The manuscript is typed by editorial staffs and is published as a beautiful book by CEO, Mr. Moongyu Park of Kyungmoon press, to them I express my sincere thanks for their cooperation.

September 2011

Department of Mathematics
Ewha Womans University
Yong Seung Cho

# CONTENTS

## Appendix A: Characteristic Class



## Appendix B: Low-Dimensional Manifold



## Appendix C: Category

PTER

1

# Riemann Surface

## 1.1 Meromorphic Function

Let $\mathbb{C}$ be the usual complex plane, and $U$ an open subset of $\mathbb{C}$.

A function $f: U \rightarrow \mathbb{C}$ is called **holomorphic** if the derivative $f'(z) = \lim_{h\to 0} \frac{f(z+h)-f(z)}{h}$ exists.

If $f(z) = u(z) + iv(z)$, then the $f(z)$ is a holomorphic function iff

$$\frac{\partial f(z)}{\partial \bar{z}} = 0$$

$$\text{iff } \frac{\partial f}{\partial x} = -i\frac{\partial f}{\partial y}$$

$$\text{iff } \frac{\partial u}{\partial x} = \frac{\partial v}{\partial y}, \quad \frac{\partial u}{\partial y} = -\frac{\partial v}{\partial x}$$

which are called the **Cauchy-Riemann differential equations**.

### Definition 1.1.1

A **Riemann Surface** is a second countable, Hausdorff topological space $\Sigma$ together with a covering of $\Sigma$ by open sets $U_\alpha$ and homeomorphisms $\varphi_\alpha$ of $U_\alpha$ onto $D_\alpha$ for each $\alpha$, where $D_\alpha$ is an open set in the complex plane $\mathbb{C}^1$, such that for all $\alpha$ and $\beta$ the map

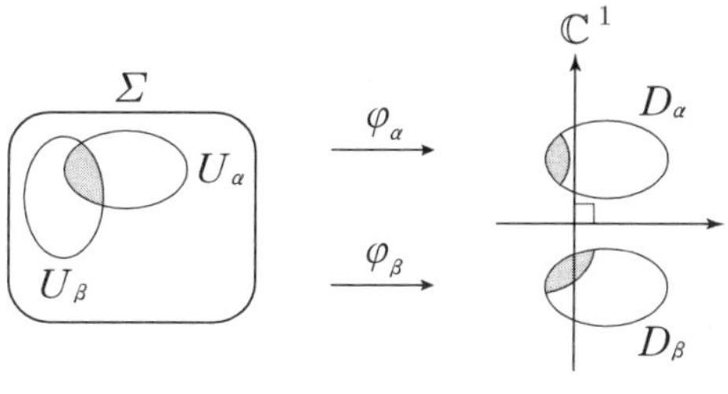


$$f_{\beta\alpha} := \varphi_\beta \circ \varphi_\alpha^{-1} : \varphi_\alpha(U_\alpha \cap U_\beta) \rightarrow \varphi_\beta(U_\alpha \cap U_\beta)$$

is holomorphic. The map $\varphi_\alpha$ allows us to introduce a **local coordinate** $z_\alpha$ in $U_\alpha$ and $f_{\beta\alpha}$ may be expressed as $z_\beta = f_{\beta\alpha}(z_\alpha)$, where $f_{\beta\alpha}$ is a holomorphic map between open subsets in $\mathbb{C}^1$.

***Remark 1.1.2***

The second countable manifolds are normal, metrizable, and paracompact. The complex plane $\mathbb{C}^1$ of complex dimension one is also called the complex line.

## Definition 1.1.3

Let $\Sigma$ be a Riemann surface, and $U$ an open subset of $\Sigma$.

A **holomorphic function** on $U$ is a function $f: U \to \mathbb{C}^1$ such that for each $\alpha$, the restriction $f_\alpha$ of $f$ to $U \cap U_\alpha$ is holomorphic in the local coordinate $z_\alpha$, i.e., for each $\alpha$

$$f_\alpha \circ (\varphi_\alpha|_{U \cap U_\alpha})^{-1} : \varphi_\alpha(U \cap U_\alpha) \to \mathbb{C}^1$$

is holomorphic.

If we are given holomorphic function $f_\alpha : U \cap U_\alpha \to \mathbb{C}$ such that $f_\alpha = f_\beta$ on $U \cap U_\alpha \cap U_\beta$ for all $\alpha, \beta$ then there is a unique holomorphic function $f: U \to \mathbb{C}^1$, whose restriction to $U \cap U_\alpha$ is $f_\alpha$ for each $\alpha$.

## Definition 1.1.4

Let $\Sigma_1$ and $\Sigma_2$ be Riemann surfaces and $f: \Sigma_1 \to \Sigma_2$ be a continuous map.

$f$ is **holomorphic** if, for each open subset $U$ of $\Sigma_2$ and each holomorphic function $g$ on $U$, $g \circ f$ is holomorphic on $f^{-1}(U)$.

For $f$ to be holomorphic it is sufficient that each $\varphi_\alpha \circ f$ is holomorphic on $f^{-1}(U_\alpha)$, where $\varphi_\alpha$ is the local coordinate of Definition (1.1.1).

### Definition 1.1.5

A **biholomorphism** of Riemann surfaces $\Sigma_1$ and $\Sigma_2$ is a holomorphic map $f: \Sigma_1 \to \Sigma_2$ such that $f^{-1}$ exists and is holomorphic.

### Examples 1.1.6

[1] The complex plane $\Sigma = \mathbb{C}^1$ and an open subset in $\mathbb{C}^1$ are Riemann surfaces.

[2] The **complex projective line** $\Sigma = \mathbb{CP}^1 = S^2$ is a Riemann surface. Indeed, we cover $\Sigma$ by $U_1 = \Sigma - \{0\}$ and $U_2 = \Sigma - \{\infty\}$. The local coordinates are $z_1$ in $U_1$ and $z_2$ in $U_2$, then $z_2 = z_1^{-1}$ in $U_1 \cap U_2$.

[3] The **complex torus** $\Sigma$ is a Riemann surface. Indeed, let $w_1$, $w_2$ be non-zero complex numbers such that $w_1/w_2$ is not real and let $L$ be the additive subgroup of $\mathbb{C}^1$ generated by $w_1$ and $w_2$. Then the **torus** $\Sigma = \mathbb{C}^1/L$ with the quotient topology, and a Riemann surface structure on $\Sigma$ is induced from that of $\mathbb{C}^1$.

[4] The nonsingular curves $\Sigma$ in $\mathbb{CP}^2$ are Riemann surfaces. The non-singular curve in $\mathbb{CP}^2$ is given by an equation $f(z_0, z_1, z_2) = 0$ where $f$ is a non-constant homogeneous polynomial.
If the degree of $f$, $\deg(f) = d$, is $d$, then the $\Sigma$ is a Riemann surface with genus $g = \dfrac{(d-1)(d-2)}{2}$.

[5] The non-constant holomorphic curves $\Sigma$ in $\mathbb{C}^2$ are given by equation $f(z_1, z_2) = 0$, where $f: \mathbb{C}^2 \to \mathbb{C}^1$ is a non-constant holomorphic function.

### Definition 1.1.7

A **meromorphic function** on a Riemann surface $\Sigma$ is a holomorphic map $f:\Sigma \to \mathbb{CP}^1 = \mathbb{C} \cup \{\infty\}$.

### Definition 1.1.8

Let $\Sigma$ be a compact Riemann surface.

(1) A **divisor** on $\Sigma$ is an element $D$ of the free abelian group generated by the points of $\Sigma$, i.e., it is a formal finite sum

$$D = n_1 P_1 + \cdots + n_k P_k,$$

where $P_1, \cdots, P_k \in \Sigma$ and $n_1, \cdots, n_k \in \mathbb{Z}$.

(2) $D \geq 0$ means that each $n_i \geq 0$,

and $D_1 \geq D_2$ means that $D_1 - D_2 \geq 0$.

(3) A **principal divisor** on $\Sigma$ is the divisor $(f)$ of a meromorphic function $f$, i.e., $(f) :=$ (zeros of $f$) $-$ (poles of $f$), where each zero and pole is counted with its multiplicity.

Since $\Sigma$ is compact and the zeros and poles of $f$ are isolated, there is only a finite number of them, so that $(f)$ is a divisor.

(4) If $\Sigma$ is not compact, these definitions have to be modified slightly. A divisor on $\Sigma$ is a (possibly infinite) formal sum $D = \Sigma n_i P_i$ which is locally finite in the sense that only a finite number of the $P_i$ are counted in any compact subset of $\Sigma$.

## 1.2 Motivation

### 1.2.1

Let $\Sigma$ be a Riemann surface and $P_i, \cdots, P_m, Q_i, \cdots, Q_n \in \Sigma$.

Is there a meromorphic function $f$ on $\Sigma$ such that $(f) = \sum_{i=1}^{m} P_i - \sum_{j=1}^{n} Q_j$?

### 1.2.2

Let a divisor $D = \Sigma n_i P_i$ be given on a Riemann surface $\Sigma$.

How many meromorphic functions $f$ are there such that $(f) + D > 0$?

### 1.2.3

Let $\Sigma$ be a Riemann surface, $P_1, \cdots, P_n$ $n$ distinct points of $\Sigma$, and let $f_i$ be meromorphic function on some neighbourhood of $P_i$ with no pole except possibly at

$$P_i \ (i = 1, \cdots, n).$$

Is there a global meromorphic function $f$ on $\Sigma$ such that $f - f_i$ is holomorphic in the neighbourhood of $P_i (i = 1, \cdots, n)$ and $f$ has no pole except $P_i, \cdots, P_n$?

## 1.3 Obstruction to the Construction of Function

Let us consider Problem (1.2.3), in the following more general formulation. Let $\Sigma$ be a Riemann surface, $\{U_\alpha\}$ an open covering of $\Sigma$, and for each $\alpha$ let $P_\alpha$ be a point of $U_\alpha$ which does not lie in any $U_\beta (\neq U_\alpha)$, and $f_\alpha$ be a meromorphic function on $U_\alpha$ with no pole except $P_\alpha$.

**Theorem 1.3.1**

(i) Let $g_{\alpha\beta} := f_\alpha - f_\beta$ on $U_\alpha \cap U_\beta$, then $g_{\alpha\beta}$ is holomorphic.

(ii) There is a global meromorphic function $f$ on $S$ such that $f - f_\alpha$ is holomorphic on $U_\alpha$ for each $\alpha$, and $f$ has no pole except the points $P_\alpha$ if and only if there is a holomorphic function $h_\alpha$ on each $U_\alpha$ such that $h_\beta - h_\alpha = g_{\alpha\beta}$ on $U_\alpha \cap U_\beta$ for all $\alpha$, $\beta$.

***Proof.*** (i) By construction $f_\alpha$ and $f_\beta$ have no pole on $U_\alpha \cap U_\beta$.

(ii) Suppose that there is a global meromorphic function $f$ on $S$ such that the restriction $f - f_\alpha$ is holomorphic on $U_\alpha$ for each $\alpha$ and $f$ has no pole except the points $P_\alpha$. We take $h_\alpha = f - f_\alpha$. The function $h_\alpha$ is holomorphic on $U_\alpha$ and

$$\begin{aligned} h_\beta - h_\alpha &= (f - f_\beta) - (f - f_\alpha) \\ &= f_\alpha - f_\beta = g_{\alpha\beta}. \end{aligned}$$

Conversely suppose that the holomorphic functions $h_\alpha$ exist on all $U_\alpha$ such that $h_\beta - h_\alpha = g_{\alpha\beta}$ on $U_\alpha \cap U_\beta$. Then we have $h_\alpha + f_\alpha = h_\beta + f_\beta$ on $U_\alpha \cap U_\beta$. Hence the $h_\alpha + f_\alpha$ can be pieced together to form a global meromorphic function $f$ on $\Sigma$ such that for each $\alpha$, $h_\alpha = f - f_\alpha$ is holomorphic on $U_\alpha$, and $f$ has no pole except the points $P_\alpha$. □

### *1.3.2 Observation*

In Theorem 1.3.1, since $g_{\alpha\beta} = f_\alpha - f_\beta$, we have

$$(\delta g)_{\alpha\beta\gamma} = g_{\beta\alpha} - g_{\alpha\gamma} + g_{\alpha\beta} = 0$$

on $U_\alpha \cap U_\beta \cap U_\gamma$ so that $g = \{g_{\alpha\beta}\}$ is a 1-cocycle.

If $h = \{h_\alpha\}$, then $(\delta h)_{\alpha\beta} = h_\beta - h_\alpha = g_{\alpha\beta}$ is a 1-coboundary.

Hence the global meromorphic function $f$ exists if and only if the 1-cocycle $g$ is a 1-coboundary if and only if $g$ defines the zero class of a certain cohomology group $H^1(X, \mathcal{O})$, where the coefficients $\mathcal{O}$ remain to be defined. The cohomology class of $g$ is the obstruction to the construction of the desired meromorphic function $f$. It will appear that the coefficient system $\mathcal{O}$ is a sheaf, namely the sheaf of germs of holomorphic functions on a Riemann surface $\Sigma$.

PTER

2

# Direct Limit

## 2.1 Direct Limit of Set

### Definition 2.1.1

A **directed set** $\Lambda$ is a set with a preorder $\leq$ (that is, a reflexive and transitive relation: $\alpha \leq \alpha$, and $\alpha \leq \beta \leq \gamma \Rightarrow \alpha \leq \gamma$).

For any $\alpha, \beta \in \Lambda$ there is $\gamma \in \Lambda$ such that $\alpha \leq \gamma$ and $\beta \leq \gamma$, we denote

$$\Lambda_1 := \{(\alpha, \beta) \in \Lambda \times \Lambda : \alpha \leq \beta\}.$$

A **direct system** of sets indexed by a directed set $\Lambda$ is a family $(U_\alpha)_{\alpha \in \Lambda}$ of sets together with, for each $(\alpha, \beta) \in \Lambda_1$, a set of maps $\rho_{\alpha\beta} : U_\alpha \to U_\beta$, satisfying

(1) for any $\alpha \in \Lambda$, $\rho_{\alpha\alpha} = \mathrm{id}_{U_\alpha}$,

(2) for any $\alpha, \beta, \gamma \in \Lambda$ if $\alpha \leq \beta \leq \gamma$ then the triangle

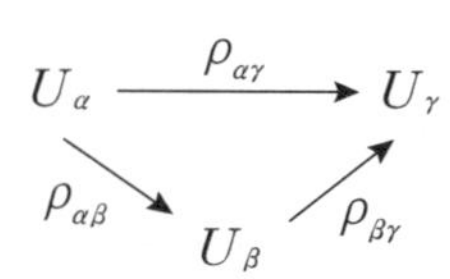


commutes, i.e., $\rho_{\alpha\gamma} = \rho_{\beta\gamma} \circ \rho_{\alpha\beta}$.

### Example 2.1.2

Given a topological space $X$, the set $\mathscr{U}$ of its open sets $U$ is directed by the relation

$$U \leq V \quad \text{iff} \quad U \supseteq V.$$

Then $\mathscr{U}$ is a direct set since $U \cap V$ is open if $U$ and $V$ are.
Given also a presheaf (See Definition (3.1.1)) $F$ on $X$, let

$$\rho_{UV} = \rho_V^U : F(U) \longrightarrow F(V)$$

be the restriction map when $U \leq V$. Then the $(F(U))_{U \in \mathscr{U}}$ with the $\rho_{UV}$ form a direct system of sets.

## Definition 2.1.3

Given a direct system $\left\{U_\alpha \xrightarrow{f_{\alpha\beta}} U_\beta\right\}_{(\alpha,\beta)\in\Lambda_1}$, we define a **target** $V$ for the system by a collection of maps $(\sigma_\alpha : U_\alpha \to V)_{\alpha\in\Lambda}$ satisfying the compatibility condition: for any $\alpha \leq \beta$,

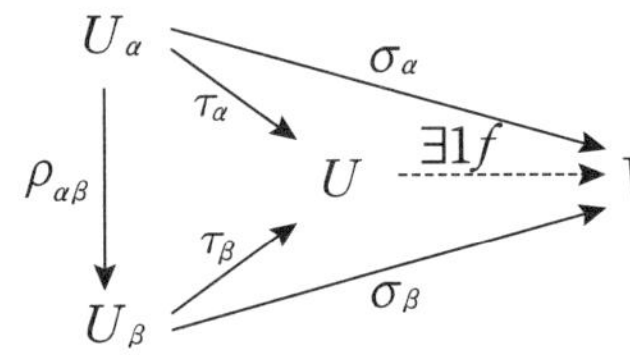


commutes, that is, $\sigma_\alpha = \sigma_\beta \circ \rho_{\alpha\beta}$.

A **direct limit** for the system $\{f_{\alpha\beta} : U_\alpha \to U_\beta \mid (\alpha,\beta)\in\Lambda_1\}$ is a target $U$, $(\tau_\alpha : U_\alpha \to U)_{\alpha\in\Lambda}$ satisfying the **universal property**:

for any target $V$ (with maps $\sigma_\alpha$ as above)

there is a unique map $f : U \to V$ such that for any $\alpha\in\Lambda$, diagram

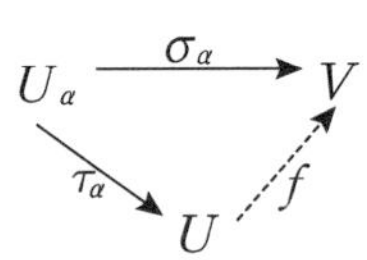


commutes, i.e., $\sigma_\alpha = f \circ \tau_\alpha$ for each $\alpha\in\Lambda$.

We denote the direct limit by $U = \varinjlim_{\alpha\in\Lambda} U_\alpha$.

From the universality, any two direct limits for a direct system are naturally isomorphic, that is, there is a bijection between them compatible with all the $\tau_\alpha$.

**Theorem 2.1.4**

Suppose that $U$, $(\tau_\alpha : U_\alpha \to U)_{\alpha \in \Lambda}$ is a target for the direct system $(U_\alpha)_{\alpha \in \Lambda}$, $(\rho_{\alpha\beta})_{(\alpha, \beta) \in \Lambda_1}$ such that

(i) for any $u \in U$, there is $\alpha (\in \Lambda)$ such that $u \in \mathrm{Im}(\tau_\alpha)$,

(ii) if $\alpha, \beta \in \Lambda$ and $u_\alpha \in U_\alpha$ and $u_\beta \in U_\beta$, then $\tau_\alpha(u_\alpha) = \tau_\beta(u_\beta)$ iff there is $\gamma \in \Lambda$ such that $\alpha \leq \gamma$, $\beta \leq \gamma$ and $\rho_{\alpha\gamma}(u_\alpha) = \rho_{\beta\gamma}(u_\beta)$.

Then $U$ is a direct limit of the direct system.

***Proof.*** Suppose that $V$, $(\sigma_\alpha : u_\alpha \to V)_{\alpha \in \Lambda}$ is another target.

Define a map $f : U \to V$ as follows:

for each $u \in U$, by condition (i) there is $\alpha \in \Lambda$ such that $\tau(u_\alpha) = u$ for some $u_\alpha \in U_\alpha$.

Then we define $f(u) = \sigma_\alpha(u_\alpha) \in V$. Also this map $f$ is unique such that $f \circ \tau_\alpha = \sigma_\alpha$ for each $\alpha \in \Lambda$.

Indeed if $g : U \to V$ such that $g \circ \tau_\alpha = \sigma_\alpha$ for each $\alpha \in \Lambda$, then for each $u \in U$, by condition (i) $u = \tau_\alpha(u_\alpha)$ for some $u_\alpha \in U_\alpha$, $f(u) = f(\tau_\alpha(u_\alpha)) = \sigma_\alpha(u_\alpha) = g(\tau_\alpha(u_\alpha)) = g(u)$.

If we choose $\beta \in \Lambda$ such that $u \in \mathrm{Im}(\tau_\beta)$, say $u = \tau_\beta(u_\beta)$, then by condition (ii) there is $\gamma \in \Lambda$ with $\rho_{\alpha\gamma}(u_\alpha) = \rho_{\beta\gamma}(u_\beta)$.

Hence

$$\begin{aligned}\sigma_\alpha(u_\alpha) &= \sigma_\gamma(\rho_{\alpha\gamma}(u_\alpha)) \\ &= \sigma_\gamma(\rho_{\beta\gamma}(u_\beta)) = \sigma_\beta(u_\beta)\end{aligned}$$

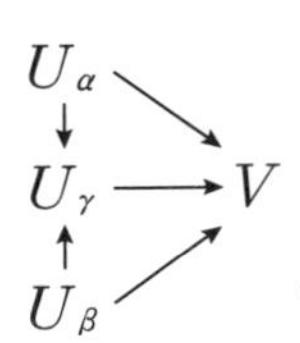


So $f$ is well-defined, and $U$ satisfies the universal property. □

## 2.2 Direct Limit of Abelian Group

In this section we will show that there are direct limits of direct systems of abelian groups. As Theorem 2.1.4 we can easily show the following Theorem 2.2.1.

**Theorem 2.2.1**

Suppose $(G_\alpha)_{\alpha \in \Lambda}$, $(\rho_{\alpha\beta} : G_\alpha \to G_\beta)_{(\alpha,\beta) \in \Lambda_1}$ is a direct system of abelian groups and $G$, $(\tau_\alpha : G_\alpha \to G)_{\alpha \in \Lambda}$ is a target satisfying

(i) for any $g \in G$, there is a $\alpha \in \Lambda$ such that $g \in \mathrm{Im}(\tau_\alpha)$

(ii) for any $\alpha$, $g_\alpha \in G_\alpha$ we have

$\tau_\alpha(g_\alpha) = 0$ iff there is $\beta$ such that $\alpha \le \beta$ and $\rho_{\alpha\beta}(g_\alpha) = 0$.

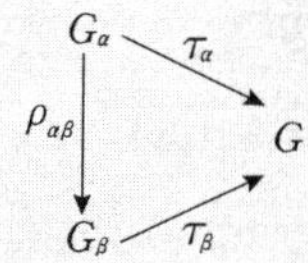


Then $G$ is a direct limit for the direct system.

**Construction 2.2.2**

Given a direct system $(G_\alpha)_{\alpha \in \Lambda}$, $(\rho_{\alpha\beta})_{(\alpha,\beta) \in \Lambda_1}$ of abelian groups, let

$$H = \oplus_{\alpha \in \Lambda} G_\alpha$$

be the abelian group (with pointwise operations) generated by the images of all the $i_\alpha$, where for $g_\alpha \in G_\alpha$ and $\beta \in \Lambda$,

$$(i_\alpha(g_\alpha))_\beta = \begin{cases} g_\alpha & \text{if } \beta = \alpha \\ 0 & \text{if } \beta \neq \alpha. \end{cases}$$

Let $H_1$ be the subgroup of $H$ generated by all the $i_\alpha(g_\alpha) - i_\beta(\rho_{\alpha\beta}(g_\alpha))$ as $(\alpha, \beta)$ runs through $\Lambda_1$ and $g_\alpha$ runs through $G_\alpha$. Put $G = H/H_1$ and $\tau_\alpha : G_\alpha \to G$ the natural maps.

**Theorem 2.2.3**

The above $G$, $(\tau_\alpha : G_\alpha \to G)_{\alpha \in \Lambda}$ is a direct limit for the direct system. Hence any direct system of abelian groups has a direct limit.

We denote the direct limit by $G = \varinjlim_{\alpha \in \Lambda} G_\alpha$.

We consider the compatibility of direct limits with exact sequences of abelian groups. Suppose we have three direct systems over the same direct set $\Lambda$, and homomorphisms

$$\left\{ U_\alpha \xrightarrow{\lambda_\alpha} V_\alpha \xrightarrow{\rho_\alpha} W_\alpha \right\}_{\alpha \in \Lambda},$$

which are exact and for $\alpha < \beta$ the diagram

$$\begin{array}{ccccc} U_\alpha & \xrightarrow{\lambda_\alpha} & V_\alpha & \xrightarrow{\rho_\alpha} & W_\alpha \\ \downarrow \rho_{\alpha\beta} & & \downarrow \rho_{\alpha\beta} & & \downarrow \rho_{\alpha\beta} \\ U_\beta & \xrightarrow{\lambda_\beta} & V_\beta & \xrightarrow{\rho_\beta} & W_\beta \end{array}$$

is commutative where the $\rho_{\alpha\beta}$ are the maps in the direct system.

Passing to the direct limit gives homomorphisms, we have

$$\varinjlim_{\alpha \in \Lambda} U_\alpha \xrightarrow{\lambda} \varinjlim_{\alpha \in \Lambda} V_\alpha \xrightarrow{\rho} \varinjlim_{\alpha \in \Lambda} W_\alpha.$$

**Theorem 2.2.4**

The limit sequence is exact.

*Remark 2.2.5*

(1) For inverse limits, the Theorem above is false.

(2) Every direct system of sets has a direct limit.

Indeed, let $(U_\alpha)_{\alpha\in\Lambda}$, $(\rho_{\alpha\beta})_{(\alpha,\beta)\in\Lambda_1}$ be a direct system of sets. Let

$$\widetilde{U}=\coprod_{\alpha\in\Lambda}U_\alpha$$

be the disjoint union of the sets $(U_\alpha)_{\alpha\in\Lambda}$. Define a relation $\sim$ on $\widetilde{U}$ by

$$u\sim v \iff [\text{if } u\in U_\alpha,\ v\in U_\beta, \text{ then there is } r\in\Lambda \text{ such that } \alpha\le r,\ \beta\le r \text{ and } \rho_{\alpha\gamma}(u)=\rho_{\beta\gamma}(v)].$$

Then $\sim$ is an equivalence relation on $\widetilde{U}$.

Let $U=\widetilde{U}/\sim$ and $\tau_\alpha : U_\alpha\to U$ be the composition

$$U_\alpha\to\widetilde{U}\to U.$$

Then $U$, $(\tau_\alpha :\ U_\alpha\to U)_{\alpha\in\Lambda}$ is the limit of the direct system.

CHAPTER

# 3 Sheaf

## 3.1 Sheaf

Let $X$ be a topological space.

### Definition 3.1.1

A **presheaf** $F$ of abelian groups on $X$ is a contravariant functor from the category of open subsets of $X$ to the category of abelian groups, that is, a family of abelian group $F(U)$ is defined on the non-empty open subsets $U$ of $X$, and homomorphisms which are called restrictions

$$\rho_V^U : F(U) \to F(V)$$

defined if $U \supseteq V$, such that

$$(*) \quad \begin{cases} \rho_U^U = \text{identity}, \\ \rho_W^U = \rho_W^V \circ \rho_V^U \text{ if } U \supseteq V \supseteq W. \end{cases}$$

***Remark 3.1.2***

Similarly we can define a presheaf of rings by that each $F(U)$ is a ring and each $\rho_V^U$ a ring homomorphism. More generally, if $\mathscr{C}$ is any category we can define a presheaf on $X$ with values in $\mathscr{C}$ as follows :

Let $\mathscr{D}$ be the category whose objects are the open subsets of $X$, and whose morphisms are the inclusion maps.

Then a **presheaf** $F$ on $X$ with values in $\mathscr{C}$ is a contravariant functor

$$F: \mathscr{D} \to \mathscr{C}.$$

The conditions $(*)$ are the conditions for $F$ to be a contravariant functor.

**Examples 3.1.3** The following $F$ in [1]~[7] are presheaves.

[1] Let $G$ be a fixed abelian group, define $F(U)=G$ for an open subset $U$ of $X$ and $\rho_V^U$ the identity on $G$ for all $U \supseteq V$.

[2] Let $F(U)=\{f: U \to \mathbb{R}$ is continuous$\}$ and $\rho_V^U(f)=f|_V$ the restriction of $f$ on $V$ if $U \supseteq V$ and $f \in F(U)$.

[3] Let $X$ be a differentiable manifold, $F(U)=\{f: U \to \mathbb{R}$ is differentiable$\}$ the additive group of all differentiable functions on $U$, $\rho$ the restrictions of functions.

[4] Let $\Sigma$ be a Riemann surface, $F(U)=\{f: U \to \mathbb{C}$ is holomorphic$\}$ the additive group of all holomorphic functions on $U$, $\rho$ the restrictions of functions.

[5] Let $\Sigma$ be a Riemann surface, $D=\Sigma n_i P_i$ a divisor on $\Sigma$.
If $U$ is an open subset of $\Sigma$, let $D|_U = \sum_{P_i \in U} n_i P_i$. Let $F(U)$ be the additive group of all holomorphic functions $f$ on $U$ such that $(f) \geq D|_U$, $\rho$ the restrictions of functions.

[6] Let $\Sigma$ be a Riemann surface, $F(U)$ the multiplicative group of all nonvanishing holomorphic functions on $U$, $\rho$ the restrictions of functions.

[7] Let $\Sigma$ be a Riemann surface, $P_1, P_2, \cdots, P_n$ n distinct points of $\Sigma$.
Let $F(U)=\prod_{p_i \in U} \mathbb{C}_i$ where each $\mathbb{C}_i=\mathbb{C}$.
If $U \supseteq V$, then we define the restriction $\rho_V^U: F(U) \to F(V)$ to be the identity on $\mathbb{C}_i$ if $P_i \in V$ and the zero map on $\mathbb{C}_i$ if $P_i \in U-V$.
This presheaf is said to be a **concentrated presheaf** on $P_1, \cdots, P_n$.

## Definition 3.1.4

Let $F$ and $G$ be presheaves of abelian groups on $X$.

A **homomorphism** $\varphi: F \to G$ is a family of homomorphisms

$$\varphi(U) : F(U) \to G(U)$$

for each open subset $U$ of $X$ such that whenever $U \supseteq V$, we have a commutative diagram

$$\begin{array}{ccc} F(U) & \xrightarrow{\varphi(U)} & G(U) \\ \rho_V^U \downarrow & & \downarrow \rho_V^U \\ F(V) & \xrightarrow{\varphi(V)} & G(V) \end{array}$$

in which the vertical arrows are the restrictions.

The homomorphism $\varphi$ is the **monomorphism (epimorphism, isomorphism)** if each $\varphi(U)$ is a monomorphism (epimorphism, isomorphism), we can define **kernel**, **image**, **exact sequence** etc., in the obvious way, and it is easily verified that the class of all presheaves of abelian groups on $X$ is an abelian category (See Appendix C).

**Examples 3.1.5**

[8] Let $\Sigma$ be a Riemann surface, $F$ the presheaf defined by a divisor $D$ as in Example [5], $G$ the presheaf defined by a divisor $D'$. If $D \geq D'$ and $f \in F(U)$ then $f \in G(U)$, hence we have a monomorphism, and $\varphi(U) : F(U) \to G(U)$ is compatible with the restriction maps. Thus $\varphi$ is a presheaf monomorphism $F \to G$.

[9] Let $\Sigma$ be a Riemann surface, $F$ the presheaf of holomorphic functions on $X$ as in Example [4], $G$ the presheaf concerntrated on $P_1, \cdots, P_n$ as in Example [7]. Define $\varphi : F \to G$ as follows; if $f \in F(U)$, then $\varphi(U)(f) = \{f(P_i)\}_{P_i \in U}$.

$\varphi(U) : F(U) \to G(U)$ is an epimorphism for all sufficiently small open sets $U$, but not necessarily for all open sets $U$, because it might happen that say $f(P_1) = f(P_2)$ for all $f$ defined at both $P_1$ and $P_2$.

[10] Let $\Sigma$ be a Riemann surface, $F$ the presheaf of holomorphic functions on $\Sigma$ as in Example [4], $G$ the multiplicative presheaf of non-vanishing holomorphic functions on $X$ as in Example [6]. Define $\varphi : F \to G$ by

$$\varphi(U) : F(U) \to G(U), \ \varphi(U)(f) : \exp 2\pi i f.$$

If $U$ is a connected open subset of $X$, the kernel of $\varphi(U)$ is isomorphic to $\mathbb{Z}$. $\varphi(U)$ is an epimorphism for all sufficiently small simply connected open sets $U$ since the **log** is single valued on $U$, i.e., for each $g \in G(U)$, let $f = \frac{1}{2\pi i} \log g \in F(U)$, then $\varphi(U)(f) = g$, but not for all open sets $U$.

## Definition 3.1.6

A **Godement sheaf** on $X$ is a presheaf $F$ satisfying the following conditions $[S_1]$ and $[S_2]$:

[$S_1$ : **monomorphism condition**] If $U$ is an open subset of $X$ and $\{U_\alpha\}_{\alpha \in \Lambda}$ is an open covering of $U$, and if $s, s' \in F(U)$ have the same image $\rho^U_{U_\alpha}(s) = \rho^U_{U_\alpha}(s') \in F(U_\alpha)$ for each $\alpha \in \Lambda$, then $s = s'$.

[$S_2$ : **gluing condition**] If $\{U_\alpha\}_{\alpha \in \Lambda}$ is an open cover of an open subset $U$ of $X$ and for each $\alpha$, $s_\alpha \in f(U_\alpha)$ are such that $s_\alpha$ and $s_\beta$ have the same image in $F(U_\alpha \cap U_\beta)$ for all $\alpha, \beta$, i.e., if $\rho^{U_\alpha}_{U_\alpha \cap U_\beta}(s_\alpha) = \rho^{U_\beta}_{U_\alpha \cap U_\beta}(s_\beta)$ for all $\alpha, \beta \in \Lambda$, then there is an element $s \in f(U)$ such that $\rho^U_{U_\alpha}(s) = s_\alpha$ for all $\alpha \in \Lambda$.

$$F(U) \to \prod_\alpha F(U_\alpha) \rightrightarrows \prod_{\alpha, \beta} F(U_\alpha \cap U_\beta)$$

**Examples 3.1.7**

[11] The presheaves of Examples [1] to [7] satisfy both $[S_1]$ and $[S_2]$, and are Godement sheaves.

[12] Suppose $F$ is a presheaf satisfying $[S_1]$. Define a presheaf $G$ by $F(U) = G(U)$ if $U \neq X$, $G(X) = F(X) \oplus A$, where $A$ is a nontrivial abelian group, $\rho_U^X : G(X) \to G(U)\,(U \neq X)$ maps $A$ to 0. Then $G$ does not satisfy $[S_1]$.

[13] Suppose $F$ is a sheaf. Define a presheaf $G$ by $G(U) = F(U)$ if $U \neq X$, and $G(X) = 0$. Then $G$ does not satisfy $[S_2]$ if $F(X) \neq 0$.

[14] Let $X = \mathbb{CP}^1$ and let $F$ be the sheaf of holomorphic functions on $X$ as in Example [4]. Let $P_1 \neq P_2$ be two points of $X$, $G$ the sheaf of holomorphic functions on $X$ which vanish at $P_1$ and $P_2$. For each open $U \subseteq X$ we have $G(U) \subseteq F(U)$. Define a presheaf $E$ by $E(U) := F(U)/G(U)$, the restriction maps $E(U) \to E(V)$ being induced by those of $F$ and $G$. Even if $F$ and $G$ are sheaves, but $E$ may not be.

Indeed, let $U_1 = X - \{P_1\}$, $U_2 = X - \{P_2\}$, $s_1$ the image in $E(U_1)$ of a function on $U_1$ which takes the value 0 at $P_2$, $s_2$ the image in $E(U_2)$ of a function on $U_2$ which takes the value 1 at $P_1$. $s_1$ and $s_2$ have the same image in $E(U_1 \cap U_2)$ since $E(U_1 \cap U_2) = 0$.

By Liouville's theorem,

$F(X)$ consists of constant functions and hence $G(X) = 0$. Thus there is no element of $E(X)$ whose images are $s_i \in E(U_i)$, $i = 1, 2$. Therefore $E$ does not satisfy the condition $[S_2]$.

## Definition 3.1.8

A **Cartan sheaf** on $X$ is a topological space $S$ together with a continuous map $\pi : S \to X$ which is called the projection and abelian group structures in each stalk $\pi^{-1}(x) \equiv S(x)$ such that

(i) $\pi$ is a local homeomorphism,

(ii) the group operations in the stalks of $S$ are continuous.

More precisely, if $E$ is the subset of $S \times S$ consisting of all $(u, v)$ such that $\pi(u) = \pi(v)$, then the map $\theta : E \rightarrow S$ defined by $\theta(u, v) = u - v$ is continuous.

***Remark 3.1.9***

(1) An equivalent formulation of the continuity condition is the following: if $x \in X$ and $u, v \in S(x)$ then given any neighbourhood $W$ of $u - v$ in $S$ there are neighbourhoods $U$ and $V$ of $u$ and $v$ respectively such that whenever $u' \in U$ and $v' \in V$ lie in the same stalk of $S$, then $u' - v' \in W$.

(2) The projection $\pi$ maps $S$ onto $X$ and the topology of $S$ induces the discrete topology on each stalk $S(x)$, indeed, since $\pi$ is a local homeomorphism, an $u \in S(x)$ has a neighbourhood $U$ which is mapped homeomorphically by $\pi$, so that $U$ meets $S(x)$ in singleton $u$.

**Examples 3.1.10**

[15] Let $X$ be a topological space, $x \in X$. We define an equivalence relation on the set of all continuous real valued functions defined in some neighbourhood of $x$ by two such functions to be equivalent if they agree on some neighbourhood of $x$. The equivalence classes are called the **germs** of continuous functions at $x$, and they form an additive abelian groups $S(x)$. Let $S$ be the disjoint union of all the groups $S(x)$ as $x$ runs through $X$. If $U$ is an open set in $X$, $f$ a continuous function on $U$, $f_x$ the equivalence class of $f$ in $S(x)$, let $V(f, U) = \{f_x \mid x \in U\}$, we take the set $\{V(f, U) \mid U :$ an open subset of $X$, $f$: a continuous function defined on $U\}$ as a base for

a topology of $S$. Then $S$ is a topological space and is a Cartan sheaf on $X$, which is called the **sheaf of germs of continuous real valued functions** on $X$.

[16] Let $S$ be the Cartan sheaf of germs of continuous functions on $\mathbb{R}^1 = X$. Let $f_1 \equiv 0$ and $f_2$ the function defined by $f_2(x) = 0$ if $x \leq 0$, $f_2(x) = x$ if $x \geq 0$. Let $V_1 = V(f_1, X)$, $V_2 = V(f_2, X)$ as in Example [15]. $V_1$ and $V_2$ meet each stalk $S(x)$ for which $x < 0$ in the same point. However at $x = 0$ the germs of $f_1$ and $f_2$ are distinct, and therefore $V_1$ and $V_2$ meet $S(0)$ in the distinct points $P_1$ and $P_2$. From the definition of topology of $S$ it follows that every neighbourhood of $P_1$ meets every neighbourhood of $P_2$, and hence $S$ is not a Hausdorff space.

[17] The **sheaf of germs of holomorphic functions** on a Riemann surface is a Hausdorff space, because of the uniqueness of analytical continuation.

[18] Let $X$ be an $n$-dimensional oriented manifold. And let
$S = \{(x, \alpha_x) \mid x \in X, \alpha_x \in H_n(X, X - x\,;\, \mathbb{Z})\}$. The set
$[(U, \alpha_U) \mid U$: open in $X$, $(U, \alpha_U) = \{(x, \alpha_x) \mid x \in U,\ \rho_x^U : H_n(X, X - U) \to H_n(X, X - x),\ \rho_x^U(\alpha_U) = \alpha_x\}]$
forms a base for a topology of $S$. Then $\pi : S \to X$, $\pi(x, \alpha_x) = x$ is a Cartan sheaf.

## 3.2 Godement Sheaf and Cartan Sheaf

### 3.2.1

Let $S$ be a Cartan sheaf on $X$, $U$ an open set in $X$.

A **section** of $S$ over $U$ is a continuous map $s : U \to S$ such that

$\pi \circ s = Id_U$ is the identity map of $U$. The set of all sections of $S$ over $U$ is an abelian group since the addition is continuous by definition, denoted by $S(U) = \Gamma(U, S)$. If $U$ and $V$ are open sets in $X$ and $U \supseteq V$, then the restriction $s \in S(U)$ to $s|_V \in S(V)$ gives a homomorphism $S(U) \to S(V)$, so that the $S(U)$ together with these restriction homomorphisms forms a presheaf on $X$ called the **presheaf of sections** of $S$. We denote this presheaf of sections by $\beta(S)$.

### 3.2.2

Conversely, if $F$ is a presheaf on $X$, we construct from it a Cartan sheaf on $X$ as follows.

For each $x \in X$, let $S(x)$ be the direct limit of the groups $F(U)$, and the restriction homomorphisms as $U$ runs through the set of all open neighbourhoods of $x$. More precisely, $S(x)$ may be described as follows: If $U$ and $V$ are open sets containig $x$, and $f \in F(U)$, $g \in F(V)$ are equivalent then there is an open neighbourhood $W$ of $x$, contained in $U \cap V$, such that $f$ and $g$ have the same image in $F(W)$. In this way we have an equivalence relation on the union of all the groups $F(U)$ for which $U$ contains $x$. The $S(x)$ is the set of equivalence classes and has an obvious abelian group structure.

For example, if $F$ is the presheaf of continuous real valued functions on $X$, then the elements of $S(x)$ are the germs of continuous functions at $x$ as in Example [15]. Let $S$ be the disjoint union of all $S(x)$, $x \in X$. We give a topology on $S$ as follows:

If $U$ is an open subset of $X$, $f \in F(U)$, let $f_x$ be the image of $f$ in $S(x)$ for each $x \in U$, and let $V(f, U) = \{f_x \mid x \in U\}$ then we take the sets $V(f, U)$, as $f$ and $U$ vary, as a base for a topology of $S$. Then $S$ is a Cartan sheaf.

The projection $\pi : S \to X$ which maps every element of $S(x)$ to $x$ is easily seen to be continuous and is a local homeomorphism.

To show that (ii) of Definition (3.1.8) is satisfied, let $u, v \in S(x)$, and let $V(f, U)$ be an open neighbourhood of $u - v$ in $S$. Then $u - v = f_x$, $u$ is the image in $S(x)$ of an element $g \in F(V)$, i.e., $u = g_x$, and $v = h_x$ for some $h \in F(W)$.

Then $g - h - f$ has a zero image in $S(x)$, and hence has a zero image in some $F(U')$ where $U' \subseteq U \cap V \cap W$. Hence $g_y - h_y = f_y$ for all $y = U'$, and therefore the neighbourhoods $V(g, U')$ and $V(h, U')$ of $u$ and $v$ respectively satisfy the condition of Definition (3.1.8) (ii).
We denote this Cartan sheaf induced by $F$ by $\alpha(F)$.

## Definition 3.2.3

If $F$ and $G$ are Cartan sheaves on $X$, then a **homomorphism** $\varphi : F \to G$ is a continuous map such that the diagram

$$\begin{array}{ccc} F & \xrightarrow{\varphi} & G \\ & {\scriptstyle\pi}\searrow \quad \swarrow{\scriptstyle\pi} & \\ & X & \end{array}$$

is commutative, where the $\pi$ are the projections, and such that for each $x \in X$ the map $\varphi(x) : F(x) \to G(x)$ is a homomorphism.
A homomorphism $\varphi$ is a **monomorphism (epimorphism)** if for each $x \in X$ $\varphi(x)$ is a monomorphism (epimorphism). If $\varphi$ is both a monomorphism and an epimorphism then $\varphi$ is called an **isomorphism**, i.e., $\varphi^{-1}$ is also a sheaf homomorphism.

### *Remark 3.2.4*

A sheaf homomorphism $\varphi$ is necessarily an open map. This is a consequence of the fact that the projections $\pi$ of the Cartan sheaves $F$

and $G$ onto $X$ are local homeomorphisms.

A subset $F'$ of a Cartan sheaf $F$ on $X$ is a **subsheaf** of $F$ if $F'$ with the induced structure is a Cartan sheaf, i.e., the inclusion $F' \to F$ is a sheaf homomorphism. From the Remark 3.2.4 $F'$ is an open subset of $F$.

We can define **kernel**, **image**, **exact sequence** etc. in the obvious way, and the class of all Cartan sheaves of abelian groups on a fixed space $X$ forms an **abelian category** (See Appendix C).

The $\alpha$ is a **functor** from the category of presheaves on $X$ to the category of Cartan sheaves on $X$, and the $\beta$ is a **functor** in the reverse direction.

**Theorem 3.2.5**

Let $X$ be a fixed topological space.

(1) If $S$ is a Cartan sheaf on $X$ there is a canonical sheaf isomorphism

$$\mu : \alpha(\beta(S)) \longrightarrow S.$$

(2) If $F$ is a presheaf on $X$ there is a canonical presheaf homomorphism.

$$\lambda : F \longrightarrow \beta(\alpha(F)).$$

(3) In (2) following statements are equivalent:

(a) $\lambda$ is an isomorphism.

(b) $F = \beta(S)$ for some Cartan sheaf $S$.

(c) $F$ is a Godement sheaf.

***Proof.*** (1) The stalk at $x$ of $\alpha(\beta(S))$ is

$$\alpha(\beta(S))(x) = \lim_{x \in U} \Gamma(U, S),$$

and an element $\bar{s}$ of this group is a germ of a section of $S$ in the neighbourhood of $x$, which is represented by a section

$s : U \to S$ by $\mu(x)(\bar{s}) = s(x)$. Then the map $\mu$ is well-defined and is a sheaf isomorphism.

(2) For each open subset $U$ of $X$ we define a map

$$\lambda(U) : F(U) \to \beta(\alpha(F))(U)$$

as follows:

Let $S = \alpha(F)$ and $f \in F(U)$, and $x \in U$. Let $f_x$ be the image of $f$ in $S(x)$. Define a section $\lambda(U)(f) : U \to S$ by $\lambda(U)(f)(x) = f_x$. Then the $\lambda(U)$ induce a homomorphism $F \to \beta(\alpha(F))$.

(3) (a)⇒(b)

Let $S = \alpha(F)$, then the $S$ is a Cartan sheaf. By (a) $\lambda : F \to \beta(S)$ is an isomorphism. Thus $F = \beta(S)$ for some Cartan sheaf $S$.

(b)⇒(c)

Suppose that $F = \beta(S)$ for some Cartan sheaf $S$.

Then the $F$ is the presheaf of sections of a Cartan sheaf $S$. The $F$ satisfies the conditions $[S_1]$ and $[S_2]$ of Definition (3.1.6). Thus $F$ is a Godement sheaf.

(c)⇒(a)

We want to show that $[S_1]$ implies that $\lambda$ is a monomorphism, and that $[S_1]$ and $[S_2]$ imply that $\lambda$ is an epimorphism.

Suppose that $F$ satisfies $[S_1]$. Let $U$ be an open subset of $X$, and let $f$ and $g$ be elements of $F(U)$ such that $\lambda(U)(f) = \lambda(U)(g)$. Then $f_x = g_x$ for each $x \in U$. Hence each $x \in U$ has an open neighbourhood $U_x \subset U$ such that $f$ and $g$ have the same image in $F(U_x)$. Hence by $[S_1]$ we have $f = g$, and each $\lambda(U)$ is a monomorphism and $\lambda$ is a presheaf monomorphism.

Suppose that $F$ satisfies $[S_1]$ and $[S_2]$. Let $S = \alpha(F)$ and $s : U \to S$ be a section. By the definition of Cartan sheaf, $s(U)$

is open in $S$ and is a union of basic open sets $V(f_\alpha, U_\alpha)$ where $U_\alpha$ is open in $U$ and $f_\alpha \in F(U_\alpha)$ for each $\alpha$. For every $x \in U_\alpha \cap U_\beta$ we have $f_{\alpha x} = f_{\beta x}$, hence each $x \in U_\alpha \cap U_\beta$ has an open neighbourhood $U_x \subset U_\alpha \cap U_\beta$ such that $f_\alpha$ and $f_\beta$ have the same image in $F(U_x)$. Applying [$S_1$] to $U_\alpha \cap U_\beta$ and covering $\{U_x\}$ we see that $f_\alpha$ and $f_\beta$ have the same image in $F(U_\alpha \cap U_\beta)$.

By applying [$S_2$] to $U$ and covering $\{U_\alpha\}$, there is an element $f \in F(U)$ whose image in $F(U_\alpha)$ is $f_\alpha$ for each $\alpha$. Hence $f_x = f_{\alpha x} = s(x)$ for each $x \in U_\alpha \cap U$. Thus $\lambda(U)$ is an epimorphism and $\lambda$ is a presheaf epimorphism. □

***Remark 3.2.6***

From Theorem 3.2.5, the functors $\alpha$ and $\beta$, and the isomorphisms $\lambda$, $\mu$ give an equivalence between the category of Godement sheaves on $X$ and the category of Cartan sheaves on $X$. This justifies us in calling both types of objects sheaves.

***Remark 3.2.7***

If $\varphi : F \to G$ is a sheaf homomorphism, then $\beta(\varphi) : \beta(F) \to \beta(G)$ is a presheaf monomorphism, and we have

(1) $\varphi$ is a sheaf monomorphism if and only if $\beta(\varphi) : \beta(F) \to \beta(G)$ is a presheaf monomorphism,
(2) $\varphi$ is a sheaf epimorphism if $\beta(\varphi)$ is a presheaf epimorphism,
(3) the converse of the second statement is false.

If $\varphi : F \to G$ is a sheaf epimorphism it does not follow that every section of $G$ is the image of $\varphi$ of the section of $F$. See the Example [9].

## 3.3 Induced Sheaf

Let $S$ be a sheaf on $X$, $S'$ a subsheaf of $S$, we define the **quotient sheaf** $S'' := S/S'$ either as a Godement sheaf or as a Cartan sheaf, and these two descriptions are equivalent.

### 3.3.1

If $S$ and $S'$ are Cartan sheaves on $X$, let $S''(x) = S(x)/S'(x)$ for each $x \in X$, and $S'' = \bigcup_{x \in X} S''(x)$. We have a canonical map $\varphi : S \to S''$ induced by $S(x) \to S''(x)$ for each $x \in X$, and give $S''$ the quotient topology, i.e., $U''$ is open in $S''$ if $\varphi^{-1}(U'')$ is open in $S$, then $S''$ is a Cartan sheaf on $X$.

### 3.3.2

If $S$ and $S'$ are Godement sheaves on $X$, we define a presheaf $T$ by $T(U) := S(U)/S'(U)$ for each open subset $U$ in $X$.

$T$ is not necessarily a Godement sheaf (Example [14]). However by (3.2.5), $S'' = \beta(\alpha(T))$ is a Godement sheaf on $X$.

### 3.3.3

Let $X$ and $Y$ be topological spaces, $f : Y \to X$ a continuous map, $G$ a sheaf on $Y$ and $F$ a sheaf on $X$.

We regard $G$ as a Godement sheaf on $X$. We define the **push down** $f_*(G)$ of $G$ by

$$f_*(G)(U) = G(f^{-1}(U))$$

for each open set $U$ in $X$, with the obvious restriction homomorphisms. Then $f_*(G)$ is a Godement sheaf on $X$.

**Example 3.3.4**

Suppose $Y$ is a closed subset of $X$, $G$ a sheaf on $Y$, and $i: Y \to X$ the inclusion of $Y$ into $X$. Then $i_*(G)$ is the extension of $G$ to $X$ as follows: at a point $x \in Y$ its stalk is the same as the stalk, $G(x)$ and at a point of $X - Y$ its stalk is 0.

We denote $i_*(G)$ by $G^X$.

***3.3.5***

Let $F$ be a Cartan sheaf on $X$, $f: Y \to X$ a continuous map.

We define the **pull-back** $f^*(F)$ of $F$ to be the subset

$$\{(u, y) \in F \times Y \mid \pi(u) = f(y)\}$$

of $F \times Y$, where $\pi: F \to X$ is the projection. Then $f^*(F)$ is a Cartan sheaf on $Y$ and we have $f^*(F)(y) = F(f(y))$ for all $y \in Y$.

**Example 3.3.6**

Let $F$ be a sheaf on $X$.

Let $Y$ be a subset of $X$, $i: Y \to X$ the inclusion. The $i^*(F)$ is the restriction of $F$ to $Y$, denote it by $F|_Y$. The stalk $F|_Y$ is just the stalk of $F$.

***3.3.7***

We may define various kinds of manifolds by using corresponding sheaves.

First we can give a slightly different definition of a Riemann surface from the usual one.

(1) A **Riemann surface** is a pair $\{X, \vartheta_X\}$ where $X$ is a second countable, Hausdorff topological space and $\vartheta_X$ is a subsheaf of the sheaf of germs of continuous complex-valued functions on $X$, such that

$\{X, \vartheta_X\}$ is locally isomorphic to $\{\mathbb{C}, \vartheta_{\mathbb{C}}\}$ where $\vartheta_{\mathbb{C}}$ is the sheaf of germs of holomorphic functions on $\mathbb{C}$. More precisely, each point $x \in X$ has an open neighbourhood $U$ and a homeomorphism $\varphi : U \to \mathbb{C}$ such that $\varphi_*(\vartheta_X|_U) \simeq \vartheta_{\mathbb{C}}$.

(2) If $(X, \vartheta_X)$ and $(Y, \vartheta_Y)$ are Riemann surfaces, then a continuous map $f : X \to Y$ is **holomorphic** if $f^*(\vartheta_Y) \subset \vartheta_X$.

(3) Also we can define **differentiable manifolds**, **complex manifolds** etc.

### *3.3.8*

Let $k$ be a fixed field. Given a topological space $X$, $F(U)$ = the set of all $k$-valued functions on $U$ is obviously a sheaf. If $k = \mathbb{R}$, then $F(U)$ = the set of all $\mathbb{R}$-valued continuous functions on $U$ is again a sheaf, since a function is continuous if and only if it is so locally. One may also consider differential manifolds or complex analytic manifolds.

If $F(U) \subset k$-valued functions and for $V \subset U$, $\rho_V^U$ is the function-theoretic restriction, then the condition $[S_1]$ is automatically satisfied, and therefore, for $F$ to be a sheaf, it suffices to verify $[S_2]$ only.

Theorem 3.3.9 is useful to construct the sheaf from a presheaf and extension of a sheaf. Theorem 3.3.10 is the gluing of sheaves.

**Theorem 3.3.9**

(a) If $F$ is a presheaf on $X$, there is a unique sheaf $\widetilde{F}$ together with the map $F \to \widetilde{F}$ such that every map $F \to G$(a sheaf on $X$) is uniquely extended to a map $\widetilde{F} \to G$.

(b) Let $B$ be an open base for a topological space $X$. Any "sheaf $F$ on $B$" is uniquely extended to a sheaf $\widetilde{F}$ on $X$ i.e., there is a sheaf $\widetilde{F}$ on $X$ such that $\widetilde{F}(U) = F(U)$ for all $U \in B$.

***Proof.*** (a) The uniqueness of $\widetilde{F}$ is clear. As for the existence: If $F$ is a presheaf of $k$-valued functions, then we simply set

$$\widetilde{F}(U)=\{f: U\to k \mid \text{there is a covering } U=\bigcup_{\alpha} U_\alpha \text{ such that } f|U_\alpha \text{ is in } F(U_\alpha)\}.$$

In a general situation, we firstly identify them if they coincide locally i.e., replace $F$ by $F'$ where $F'(U)=F(U)/\sim$, where $\phi \sim \psi$ if there is an open covering $U=\bigcup_{\alpha} U_\alpha$ such that $\rho^U_{U_\alpha}(\phi)=\rho^U_{U_\alpha}(\psi)$ for all $\alpha$. If $\mathscr{U}=\{U_\alpha\}$, $U=\bigcup_{\alpha} U_\alpha$, is an open covering of $U$ we set

$$k_{\mathscr{U}}=\{(\phi_\alpha)\mid \phi_\alpha\in F'(U_\alpha) \text{ and } \rho^{U_\alpha}_{U_\alpha\cap U_\beta}\phi_\alpha=\rho^{U_\beta}_{U_\alpha\cap U_\beta}\phi_\beta \text{ for all } \alpha,\beta\},$$

$$\text{and take } \widetilde{F}(U)=\varinjlim_{\mathscr{U}} k_{\mathscr{U}}.$$

(b) For each open set $U\subset X$ we set $B_U=\{U_\alpha\in B\mid U_\alpha\subset U\}$ which is an open base of $U$. Since $\widetilde{F}$ is a sheaf we must have that

$$\widetilde{F}(U)=\{(\phi_\alpha)\mid U_\alpha\in B_U,\ \phi_\alpha\in\widetilde{F}(U_\alpha)=F(U_\alpha),\ \rho^{U_\alpha}_{U_\alpha\cap U_\beta}\phi_\alpha=\rho^{U_\beta}_{U_\alpha\cap U_\beta}\phi_\beta\},$$

and therefore the uniqueness as well as the existence follows. One may observe that when $F$ is of $k$-valued function, then $\widetilde{F}(U)=\{f: U\to k\mid f|_{U_\alpha} \text{ is in } F(U_\alpha) \text{ for all } U_\alpha\in B_U\}$. □

**Theorem 3.3.10**

Let $X = \bigcup_\alpha U_\alpha$ be an open covering and suppose that we are given sheaves $(U_\alpha, F_\alpha)$ and an isomorphism

$$\phi_{\alpha, \beta} : F_\alpha | U_\alpha \cap U_\beta \xrightarrow{\sim} F_\beta | U_\alpha \cap U_\beta$$

such that

$$\phi_{\alpha, \alpha} = id_{F_\alpha} \text{ and } \phi_{\beta, \gamma} \cdot \phi_{\alpha, \beta} = \phi_{\alpha, \gamma} \text{ on } U_\alpha \cap U_\beta \cap U_\gamma.$$

Then there exists a unique sheaf $F$ on $X$ together with an isomorphism

$$\eta_\alpha : F | U_\alpha \xrightarrow{\sim} F_\alpha$$

such that $\phi_{\alpha, \beta} = \eta_\beta \cdot \eta_\alpha^{-1}$ for all $\alpha, \beta$.

***Proof.*** Define $F$ by

$$\begin{aligned} F(V) = \{(f_\alpha) | f_\alpha \in F_\alpha (V \cap U_\alpha), \\ \phi_{\alpha, \beta}(f_\alpha | V \cap U_\alpha \cap U_\beta) = f_\beta | V \cap U_\alpha \cap U_\beta\}. \end{aligned}$$

Then $F$ is the required unique sheaf. □

TER

4

# Geometric Space

In this Chapter, we study spaces equipped with a sheaf of rings, and the geometric spaces where the stalks are all local rings, and morphisms between geometric spaces specialize to the appropriate kinds of maps between several types of manifolds.

We then consider sheaves of modules over a ringed space, which generalize the idea of vector bundles, and globalize the idea of a module over a ring. The module constructions of direct sum and product, tensor product and module of homomorphisms also globalize to these sheaves, with universal properties. Change of base space by a morphism of ringed spaces gives rise to direct and inverse image functors.

Throughout this Chapter the word 'ring' will mean commutative ring with a one, and ring morphisms are required to preserve ones.

## 4.1 Ringed Space

### 4.1.1

Let $R$ be a ring. An $R$-algebra is a ring $S$ together with a ring morphism $\alpha : R \to S$ called the structure map. A morphism of $R$-algebras $\beta : S \to S'$ is a ring morphism making the triangle

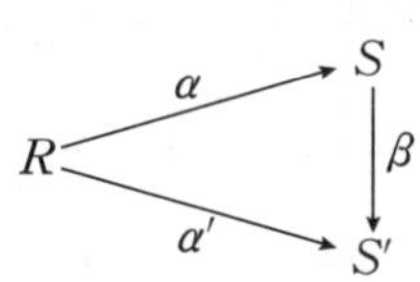


commute. Hence we have a category $R$-alg of all $R$-algebras.

For instance, for any ring $S$ there is a unique ring morphism $\mathbb{Z} \to S$ (taking $n \mapsto n1_S$) and any ring morphism $S \to S'$ makes the triangle

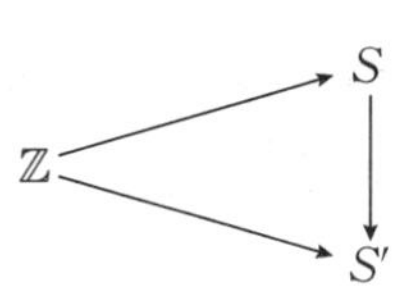

commute. Hence any ring is a $\mathbb{Z}$-algebra in a unique way, and indeed the category $\mathbb{Z}$-alg is isomorphic to the category of rings. If $R$ is a field, such as $\mathbb{R}$, $\mathbb{C}$ or the $p$-adic numbers $\mathbb{Q}_p$ for a prime number $p \in \mathbb{N}$, then an $R$-algebra is an $R$-vector space together with an $R$-linear multiplication.

## Definition 4.1.2

If $R$ is a ring, a ringed space over $R$ is a pair$(X, \mathscr{O}_X)$ where $X$ is a topological space and $\mathscr{O}_X$ is a sheaf of $R$-algebras on $X$.

Of course a sheaf of $R$-algebras is a presheaf of $R$-algebras which satisfies the sheaf conditions $[S_1]$ and $[S_2]$ a morphism of sheaves of $R$-algebras is defined in the obvious way. We can regard $\mathscr{O}_X$ as a sheaf of abelian groups (forgetting the multiplicative structure).

By abuse of language, we often say '$X$ is a ringed space over $R$' and call $\mathscr{O}_X$ the structure sheaf. If no mention is made of $R$, it is often assumed that $R = \mathbb{Z}$, so that after (4.1.2) $\mathscr{O}_X$ is just a sheaf of rings.

It is easy to see that the direct limit of a direct system of $R$-algebras and morphisms has a natural $R$-algebra structure; in particular the stalks $\mathscr{O}_{X,x}$ for $x \in X$ are $R$-algebras.

**Theorem 4.1.3**

Suppose that $(E, p)$ is a sheaf space over $X$ such that each fibre $p^{-1}(x)(x \in X)$ has a given $R$-algebra structure. Then $E$ is the sheaf space of a sheaf of $R$-algebras iff equivalently

(1) for each open $U$ in $X$, $\Gamma(U, E)$ is an $R$-algebra under pointwise addition and multiplication of functions;

(2) let $E \pi E = \{(e, e') \in E \times E \mid p(e) = p(e')\}$; then the maps $(e, e') \mapsto$

$e+e'$ and $(e,e')\mapsto ee'$ (ring operations in $p^{-1}(p(e))$) are continuous: $E\pi E\rightarrow E$, and for every $r\in R$ the map $E\rightarrow E$: $e\mapsto re$ is continuous.

***Proof.*** We consider the diagram

$$\begin{array}{ccc} E\pi E & \longrightarrow & E \\ & {\scriptstyle p}\searrow \quad \swarrow{\scriptstyle p} & \\ & X & \end{array}.$$

If $f,\ g\in\Gamma(U,\ E)$, then

$$f+g,\ \ f\cdot g,\ \ U\rightarrow E$$

can be written as the composition

$$\begin{aligned} &U\rightarrow E\pi E\rightarrow E,\\ &x\mapsto(f(x),\ g(x))\mapsto f(x)+g(x),\\ &\qquad\qquad\qquad\qquad\quad f(x)\cdot g(x) \end{aligned}$$

and so is continuous.

Therefore $f+g,\ f\cdot g\in\Gamma(U,\ E)$.

Similarly, for any $r\in R$, $rf\in\Gamma(U,\ E)$. □

**Corollary 4.1.4** If $X$ and $Y$ are ringed spaces over a ring $R$, and $\phi$: $X\rightarrow Y$ is a continuous map, then the sheaves $\phi^{*}\mathcal{O}_Y$ and $\phi_{*}\mathcal{O}_X$ are sheaves of $R$-algebras on $X$ and $Y$ respectively, and the natural morphisms $\mathcal{O}_Y\rightarrow\phi_{*}\phi^{*}\mathcal{O}_Y$, $\phi^{*}\phi_{*}\mathcal{O}_X\rightarrow\mathcal{O}_X$ are morphisms of sheaves of $R$-algebras. □

**Examples 4.1.5**

[1] For any topological space $X$, letting the structure sheaf $\mathcal{O}_X=\mathbb{Z}_X$, the constant sheaf, makes $X$ into a ringed space over $\mathbb{Z}$.

[2] Let $R$ be a topological ring, that is a ring endowed with a topology such that the ring operations (of subtraction and multiplication) are continuous; for example $\mathbb{R}$, $\mathbb{C}$, the $p$-adic numbers $\mathbb{Z}_p$, or any ring with the discrete topology. For any topological space $X$, let $\mathscr{O}_X = C^R$ be the sheaf of continuous functions on $X$ with values in $R$. Then $(X, \mathscr{O}_X)$ is a ringed space over $R$; the $R$-algebra structures $R \to \Gamma(U, \mathscr{O}_X)$ for $U$ open in $X$ are given by the constant functions. For $R = \mathbb{Z}$ with the discrete topology, this gives example [1].

[3] Let $X$ be a banach space over $R$ (for instance $R^n$ for $n \in \mathbb{N}$) and let $\mathscr{O}_X = C^r$ be the sheaf of $r$-times continuously differentiable $R$-valued functions on $X$. Then $(X, \mathscr{O}_X)$ is a ringed space over $R$. Note that for $U$ open in $X$, the sheaf $C^r$ defined on $U$ as in loc. cit. is just $\mathscr{O}_X | U$.

[4] Let $K$ be a complete non-discrete valued field, such as $\mathbb{R}, \mathbb{C}$ or the $p$-adic numbers $\mathbb{Q}_p$; let $X$ be a banach space over $K$. Then the structure sheaf $\mathscr{O}_X = C^\omega$ of all $K$-valued analytic functions on $X$ makes $X$ into a ringed space over $K$.

## Definition 4.1.6

Let $R$ be a ring. A **morphism**

$$\Phi : (X, \mathscr{O}_X) \to (Y, \mathscr{O}_Y)$$

of ringed spaces over $R$ is given by a continuous map $\phi : X \to Y$ of the underlying topological spaces together with a $\phi$-morphism of sheaves of $R$-algebras $\mathscr{O}_Y \to \mathscr{O}_X$, this may be regarded as a morphism $\mathscr{O}_Y \to \phi_* \mathscr{O}_X$ of sheaves of $R$-algebras on $Y$.

Given morphisms $(X, \mathscr{O}_X) \to (Y, \mathscr{O}_Y) \to (Z, \mathscr{O}_Z)$ of ringed spaces over $R$, we get a composite morphism $(X, \mathscr{O}_X) \to (Z, \mathscr{O}_Z)$ by composing the underlying continuous maps $X \xrightarrow{\phi} Y \xrightarrow{\psi} Z$ and combining $\mathscr{O}_Y \to \phi_* \mathscr{O}_X$ and $\mathscr{O}_Z \to \psi_* \mathscr{O}_Y$ to get $\mathscr{O}_Z \to \psi_* \mathscr{O}_Y \to \psi_* \phi_* \mathscr{O}_X = (\psi \circ \phi)_* \mathscr{O}_X$.

Hence we have defined a category of all ringed spaces over $R$.

**Theorem 4.1.7**

$\Phi : (X, \mathscr{O}_X) \to (Y, \mathscr{O}_Y)$ is an isomorphism of ringed spaces over $R$ iff $\Phi$ is invertible (that is, there is $\Psi$ such that $\Phi \circ \Psi = id$, $\Psi \circ \Phi = id$) iff the underlying map $\phi : X \to Y$ is a homeomorphism and the map $\mathscr{O}_Y \to \phi_* \mathscr{O}_X$ is an isomorphism of sheaves of $R$-algebras.

***Proof.*** The first equivalence is by definition, and the second follows easily. □

**Example 4.1.8**

Let $(X, \mathscr{O}_X)$ and $(Y, \mathscr{O}_Y)$ be ringed spaces of the kind described in Example 4.1.5[2] (resp. [3], [4]), and let $\phi : X \to Y$ be a continuous (resp. differentiable, resp. analytic) map. Then there is a morphism of ringed spaces $(X, \mathscr{O}_X) \to (Y, \mathscr{O}_Y)$ with underlying continuous map $\phi$, given by 'composition with $\phi$' : the $\phi$-morphism $\mathscr{O}_Y \to \mathscr{O}_X$ is defined by

$$\Gamma(V, \mathscr{O}_Y) \to \Gamma(\phi^{-1} V, \mathscr{O}_X) \ : \ s \mapsto s \circ \phi$$

for $V$ open in $Y$.

**Theorem 4.1.9**

If $\phi : X \to Y$ is a continuous map, and $G$ is a presheaf on $Y$, then the sheaf $\phi^* G$ with the morphism $n : G \to \phi_* \phi^* G$ has the universal property that for any sheaf $F$ on $X$, the map

$$\mathrm{Hom}_X(\phi^* G, F) \to \mathrm{Hom}_Y(G, \phi_* F) : f \mapsto (\phi_* f) \circ n$$

is bijective.

Furthermore this property characterizes $\phi^* G$ uniquely up to isomorphism in $\mathrm{Shv}/X$.

Thus we have natural isomorphisms:

$$\mathrm{Hom}_X(\phi^* G, F) \xrightarrow{\sim} \mathrm{Hom}_\phi(G, F) \xrightarrow{\sim} \mathrm{Hom}_Y(G, \phi_* F)$$

whenever $F$ is a sheaf on $X$ and $G$ a presheaf on $Y$.

***Proof.*** Given $f : \phi^* G \to F$ we obtain $G \xrightarrow{n} \phi_* \phi^* G \xrightarrow{\phi_* f} \phi_* F$. The inverse map is : given $g : G \to \phi_* F$, compose $\phi^* G \xrightarrow{\phi^* g} \phi^* \phi_* F \to F$.

That these procedures are inverse may be seen by considering the procedures:

given $f : \phi^* G \to F$ we get a diagram for $V$ open in $Y$, and the dotted map gives $G \to \phi_* F$;

$$\begin{array}{ccc} \Gamma(\phi^{-1}V, \phi^* G) & \longleftarrow & G(V) \\ \downarrow & & \vdots \\ \Gamma(\phi^{-1}V, F) & = & \Gamma(V, \phi_* F) \end{array}$$

given $g : G \to \phi_* F$ we get $\forall x \in X$ : and the dotted maps give $\phi^* G \to F$(and determine it uniquely since $\phi^* G$ and $F$ are both sheaves).

$$\begin{array}{ccc} (\phi^* G)_X & \xrightarrow{\sim} & G_{\phi(X)} \\ \vdots & & \downarrow \\ F_X & \longleftarrow & (\phi_* F)_{\phi(X)} \end{array}$$

For the uniqueness, if also $\bar{n} : G \to \phi_* \bar{G}$ has the property, then $n$, $\bar{n}$ factor through each other : say

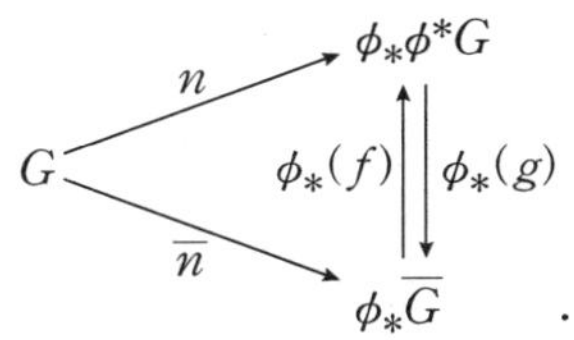

Then $g \circ f : \overline{G} \to \overline{G}$ has $\phi_*(g \circ f) \circ \overline{n} = \phi_*(id_{\overline{G}}) \circ \overline{n}$ and so $g \circ f = id_{\overline{G}}$.

Similarly $f \circ g = id_{\phi^* G}$, and so $\phi^* G \xrightarrow{\sim} \overline{G}$. □

**Theorem 4.1.10**

Any morphism $\Phi : (X, \mathscr{O}_X) \to (Y, \mathscr{O}_Y)$ of ringed spaces over $R$ can be factored uniquely as

$$(X, \mathscr{O}_X) \xrightarrow{\Psi} (X, \phi^* \mathscr{O}_Y) \xrightarrow{\Theta} (Y, \mathscr{O}_Y)$$

with $\Psi$ having an underlying continuous map $id_X$ and $\Theta$ having as morphism of sheaves of $R$-algebras the natural morphism $\mathscr{O}_Y \to \phi_* \phi^* \mathscr{O}_Y$ (4.1.9).

Moreover this factorization has the universal property that if $(X, \mathscr{O}'_X)$ is another ringed space over $R$ through which $\Phi$ factors as

$$(X, \mathscr{O}_X) \xrightarrow{\Psi'} (X, \mathscr{O}'_X) \xrightarrow{\Theta'} (Y, \mathscr{O}_Y)$$

with $\Psi'$ having $id_X$ as an underlying continuous map, then there is a unique morphism $\phi^* \mathscr{O}_Y \to \mathscr{O}'_X$ making the triangle

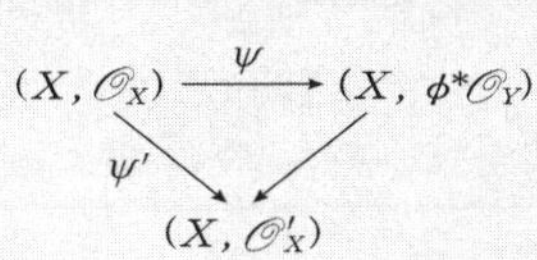

commute.

**Proof.** The existence and uniqueness of $\Psi$ follow from (4.1.9) the given morphism $\mathcal{O}_Y \to \phi_*\mathcal{O}_X$ associated with $\Phi$ corresponds uniquely to a morphism $\phi^*\mathcal{O}_Y \to \mathcal{O}_X$. The universal property of $\Psi$ also follows from (4.1.9) using the bijection

$$\mathrm{Hom}\,(\mathcal{O}_Y,\ \phi_*\mathcal{O}'_X) \xrightarrow{\sim} \mathrm{Hom}(\phi^*\mathcal{O}_Y,\ \mathcal{O}'_X).$$

In each use of (4.1.9) we must make the easy observation that the bijection restricts to a bijection between sets of morphisms of sheaves of $R$-algebras. □

**Theorem 4.1.11**

Any morphism $\Phi: (X, \mathcal{O}_X) \to (Y, \mathcal{O}_Y)$ of ringed spaces over $R$ can be factored uniquely as

$$(X, \mathcal{O}_X) \xrightarrow{\Psi} (Y, \mathcal{O}'_Y) \xrightarrow{\Theta} (Y, \mathcal{O}_Y)$$

with $\Theta$ having as an underlying continuous map $id_Y$ and a morphism of sheaves $\mathcal{O}_Y \to id_*\mathcal{O}'_Y = \mathcal{O}'_Y$ a sheaf epimorphism, while the sheaf morphism $\mathcal{O}'_Y \to \phi_*\mathcal{O}_X$ associated with $\Psi$ is a monomorphism.

$\Theta$ is then characterized by the subsheaf (of abelian groups) $I = \mathrm{Ker}(\mathcal{O}_Y \to \mathcal{O}'_Y)$ of $\mathcal{O}_Y$, which is a sheaf of ideals in $\mathcal{O}_Y$ : that is, equivalently

for each $y \in Y$, $I_y$ is an ideal in $\mathcal{O}_{Y,y}$

and $V$ open in $Y$, $\Gamma(V, I)$ is an ideal in $\Gamma(V, \mathcal{O}_Y)$.

**Proof.** Set the sheaf image $\mathcal{O}'_Y = \mathrm{SIm}(\mathcal{O}_Y \to \phi_*\mathcal{O}_X)$; this is easily seen to be a sheaf of $R$-algebras with the desired properties, since

$\mathscr{O}_Y \to \phi_* \mathscr{O}_X$ factors as $\mathscr{O}_Y \to \mathscr{O}'_Y \to \phi_* \mathscr{O}_X$. The exact sequence of sheaves of abelian groups

$$0 \to I \xrightarrow{f} \mathscr{O}_Y \xrightarrow{g} \mathscr{O}'_Y \to 0$$

shows that $I = \mathrm{Ker}(g)$ and $\mathscr{O}'_Y = \mathrm{SCok}(f)$ determine each other, and the stalk sequence over $y \in Y$ shows that $I_y$ is an ideal of $\mathscr{O}_{Y,y}$. The equivalence of the second condition follows from the sheaf space interpretation (4.1.3) □

**Theorem 4.1.12**

If $\Theta$: $(X, \mathscr{O}_X) \to (Y, \mathscr{O}_Y)$ is a morphism of ringed spaces over $R$ such that the underlying map $\Theta : X \to Y$ is injective and the morphism $\mathscr{O}_Y \to \theta_* \mathscr{O}_X$ is a sheaf epimorphism, then $\Theta$ is a monomorphism in the category of ringed spaces over $R$.

***Proof.*** We must show that given a diagram

$$(Z, \mathscr{O}_Z) \underset{\Psi}{\overset{\Phi}{\rightrightarrows}} (X, \mathscr{O}_X) \xrightarrow{\Theta} (Y, \mathscr{O}_Y)$$

such that $\Theta \circ \Phi = \Theta \circ \Psi$, we can deduce $\Phi = \Psi$. The diagram of underlying continuous maps is

$$Z \underset{\psi}{\overset{\phi}{\rightrightarrows}} X \xrightarrow{\theta} Y$$

and since $\theta$ is injective, $\phi = \psi$. The two composites

$$\mathscr{O}_Y \to (\theta_* \mathscr{O}_X) \rightrightarrows (\theta \circ \phi)_* \mathscr{O}_Z$$

are equal by hypothesis. Applying (4.1.9) we get a diagram of sheaves on $Z$

$$(\theta \circ \phi)^* \mathscr{O}_Y = \phi^* \theta^* \mathscr{O}_Y \to \phi^* \mathscr{O}_X \rightrightarrows \mathscr{O}_Z$$

in which the composites are equal too. The stalk sequence over $z \in Z$ is

$$\begin{array}{ccc}
((\theta \circ \phi)^* \mathscr{O}_Y)_z & \to & (\phi^* \mathscr{O}_X)_z \rightrightarrows \mathscr{O}_{Z,\, z} \\
\wr\downarrow & & \wr\downarrow \\
\mathscr{O}_{Y,\; \theta(\phi(z))} & \xrightarrow{\alpha} & \mathscr{O}_{X,\; \phi(z)}
\end{array}$$

and by hypothesis $\alpha$ is surjective. Hence the two maps $\phi^* \mathscr{O}_X \rightrightarrows \mathscr{O}_Z$ are equal, and by (4.1.9) so are the maps $\mathscr{O}_X \rightrightarrows \phi_* \mathscr{O}_Z$ defining $\Phi$ and $\Psi$. Hence $\Phi = \Psi$. □

***Remark 4.1.13***

Any morphism $\Phi : (X, \mathscr{O}_X) \to (Y, \mathscr{O}_Y)$ of ringed spaces over $R$ can be factored as

$$\begin{array}{ccccc}
 & \nearrow & (Y,\; \mathscr{O}'_Y) & \overset{\Theta}{\searrow} & \\
(X,\; \mathscr{O}_X) & & & & (Y,\; \mathscr{O}_Y) \\
 & \underset{\Psi}{\searrow} & (X,\; \phi^* \mathscr{O}_Y) & \nearrow & 
\end{array}$$

such that $\Theta$ is a monomorphism with an underlying continuous map $id_Y$ and $\Psi$ has an underlying continuous map $id_X$ and is final among those ringed space structures on $X$ through which $\Phi$ factors.

**Example 4.1.14**

Let $X = \{x\}$ be a one point space and $Y = \mathbb{R}^2$ be the real plane. Give $X$ and $Y$ the structure sheaves $\mathscr{O}_X$, $\mathscr{O}_Y$ of continuous $R$-valued functions (4.1.5[2]), and let $\phi : X \to Y$ map $x$ to the point $y_0 \in Y$. The associated morphism of ringed spaces over $R$ (4.1.8) has stalk map

$$\mathscr{O}_{Y,\, y_0} \to (\phi_* \mathscr{O}_X)_{y_0} \cong \mathscr{O}_{X,\, x} \cong R$$

assigning to the germ of a function $f$ its value $f(y_0)$ at $y_0$.

The sheaf $\mathscr{O}'_Y$ of (4.1.11) has stalks

$$\mathscr{O}'_{Y,\,y} = \begin{cases} R & \text{if } y = y_0 \\ \{0\} & \text{if } y \neq y_0 \end{cases}$$

while the sheaf of ideals $I$ of (4.1.11) has stalks

$$I_y = \mathscr{O}_{Y,\,y} \text{ if } y \neq y_0$$

and

$$I_{y_0} = \{g \in \mathscr{O}_{Y,\,y_0} \mid g \text{ is the germ of a function } f \text{ such that } f(y_0) = 0\}.$$

The sheaf $\phi^* \mathscr{O}_Y$ on $X$ of (4.1.10) has a stalk

$$(\phi^* \mathscr{O}_Y)_x = \mathscr{O}_{Y,\,y_0},$$

and the map $\mathscr{O}_{Y,\,y_0} \to \mathscr{O}_{X,\,x} \cong R$ is again given by taking the value at $y_0$.

□

## 4.2 Prime Spectrum of Ring

### *4.2.1*

Let $R$ be a commutative ring with one. We can associate with $R$ a ringed space $(\mathrm{Spec}R,\ \mathscr{O})$ over $R$ (or over $\mathbb{Z}$) as follows.

An ideal $p$ of $R$ is called prime iff the ring $R/p$ is integral (= has no zero divisors and is not the zero ring). Let $\mathrm{Spec}R$ be the set of prime ideals of $R$, and for any ideal $a$ or $R$ let

$$V(a) = \{p \in \mathrm{Spec}R \mid p \supseteq a\}.$$

Then the $V(a)$, as $a$ runs through all ideals of $R$, form the closed sets of a topology: this follows from the easy observations:

**Lemma 4.2.2**

(i) $V(\{0\}) = \mathrm{Spec}R$; $V(R) = \phi$.

(ii) $V(\Sigma_i a_i) = \cap_i V(a_i)$.

(iii) $V(ab) = V(a) \cup V(b)$.

***Proof.*** The only difficulty is $V(ab) \subseteq V(a) \cup V(b)$; but if $p \in (\mathrm{Spec}R) - (V(a) \cup V(b))$ then $\exists f \in a - p$ and $\exists g \in b - p$ so that $fg \notin p$ but $fg \in ab$; therefore $p \not\supseteq ab$. □

### *4.2.3*

For $f \in R$, let

$$D(f) = \{ p \in \mathrm{Spec}R \mid f \notin p \} = \mathrm{Spec}R - V(Rf).$$

Then the $D(f)$ for $f \in R$ form a basis for the topology on $X = \mathrm{Spec}R$; for if $p \in X - V(a)$ then picking $f \in a - p$, we have $p \in D(f) \subseteq X - V(a)$.

Also $D(f) \cap D(g) = D(fg)$, so that the basis $\{D(f) \mid f \in R\}$ is closed under finite intersections.

**Examples 4.2.4**

[1] If $k$ is any field, Spec$k$ has just one point, $\{0\}$.

[2] Let $k$ be an algebraically closed field (e.g. $\mathbb{C}$), and let $R = k[x]$ be the polynomial ring in one variable. The only prime ideals of $R$ are $\{0\}$ and the $R(x - \alpha)$ for $\alpha \in k$. Hence $\mathrm{Spec}(k[x]) = \{\{0\}\} \cup M$ where $M \xrightarrow{bij} k$; the topology induced on $M$ has the finite sets as its closed sets, and in Spec$R$ the closure of the point $\{0\}$ is the whole space.

[3] If $k$ is an algebraically closed field, then

$$\mathrm{Spec}k[x_1, \cdots, x_n] \supseteq M$$

where $M$ is the set of closed points (maximal ideals), and by the

Weak Nullstellensatz $M \xrightarrow{bij.} k^n$, since any maximal ideal is of the form $(x_1 - \alpha_1, x_2 - \alpha_2, \cdots, x_n - \alpha_n)$ for some $\alpha_i \in k$ $(1 \leq i \leq n)$.

The topology induced on $k$ is the Zariski topology, having as its closed sets all the sets of common zeros of a collection of polynomials: that is, all sets of the form

$$\{(\alpha_1, \cdots, \alpha_n) \in k^n \mid 1 \leq i \leq m, \ \ f_i(\alpha_1, \alpha_2, \cdots, \alpha_n) = 0\}$$

where $f_1, \cdots, f_m \in k[x_1, \cdots, x_n]$.

[4] Spec$Z$ has one closed point for each prime number $p \in \mathbb{N}$, and one other point.

### *4.2.5*

To each basic open set of Spec$R$ we can assign a ring:

$$D(f) \mapsto R_f,$$

where $R_f$ is the ring of fractions of $R$ with respect to the multiplicative system $\{1, f, f^2, \cdots\}$, and to any containment $D(f) \supseteq D(g)$ a restriction ring morphism $R_f \to R_g$; namely

$$D(f) \supseteq D(g) \Leftrightarrow \exists n \in N, \ a \in R \text{ such that } g^n = af;$$

then define

$$R_f \to R_g : \frac{r}{f^m} \mapsto \frac{a^m r}{g^{nm}}.$$

In particular, if $D(f) = D(g)$, then $R_f \xrightarrow{\sim} R_g$, so that the assignment $D(f) \mapsto R_f$ is 'well-defined'.

Since $D(f) \cap D(g) = D(fg)$, if $D(f) = \cup_{\lambda \in \Lambda} D(f_\lambda)$ is a cover of a basic open set by basic open sets, we have the sequence

$$R_f \to \prod_{\lambda\in\Lambda} R_{f_\lambda} \rightrightarrows \prod_{(\lambda,\mu)\in\Lambda\times\Lambda} R_{f_\lambda f_\mu}$$

and we can check by commutative algebra that this sequence is an equaliser.

Hence we have the data for a 'sheaf defined only on the open sets of a base', and we need the following sheaf-theoretic Lemma.

**Lemma 4.2.6**

Suppose that $X$ is a topological space and $\mathscr{U}$ is a basis for the topology which is closed under finite intersections. Let $F$ be the data of a presheaf (sets of sections and restriction maps) given only for open sets of the basis, which satisfies the condition that whenever $U = \cup_{\lambda\in\Lambda} U_\lambda$ with $U \in \mathscr{U}$ and $\forall\lambda\in\Lambda$, $U_\lambda\in\mathscr{U}$, the sequence

$$F(U) \to \prod_{\lambda\in\Lambda} F(U_\lambda) \rightrightarrows \prod_{(\lambda,\mu)\in\Lambda\times\Lambda} F(U_\lambda\cap U_\mu)$$

is an equaliser diagram (maps as in 3.1.6)

Then there is a sheaf $G$ on $X$, unique up to isomorphism, such that

$$\forall U\in\mathscr{U} \qquad \Gamma(U, G) = F(U)$$

and $\forall U, V\in\mathscr{U}$, if $U\supseteq V$ the restriction maps $F(U)\to F(V)$ and $\Gamma(U, G)\to\Gamma(V, G)$ agree.

***Sketch of proof***

Step 1 : If such a $G$ exists, it must have stalks

$$G_x = \varinjlim_{x\in U\in\mathscr{U}} F(U)$$

so that the underlying set of the sheaf space $LG$ is determined.

Step 2 : The topology on $LG$ is also determined, since the $\hat{s}[U]$ with $U \in \mathscr{U}$ and $s \in F(U)$ must form a basis for it. Hence the sheaf $G$ is unique if it exists.

Step 3 : Construct $G$ by forming the stalks $G_x = \varinjlim_{x \in U \in \mathscr{U}} F(U)$ and thus a sheaf space $E = \coprod_{x \in X} G_x$, and check that $F(U) \xrightarrow{\sim} \Gamma(U, E)$ for $U \in \mathscr{U}$. □

An alternative construction is to set, for $V$ open in $X$,

$$\Gamma(V, G) = \varprojlim_{V \supset U \in \mathscr{U}} F(U).$$

### 4.2.7

Applying Lemma 4.2.6 to the situation of (4.2.5) we see that $X = \operatorname{Spec} R$ has on it a sheaf of rings, denoted by $\mathscr{O}$ or $\mathscr{O}_X$, with the property that for $f \in R$

$$\Gamma(D(f), \mathscr{O}) = R_f.$$

In particular, putting $f = 1$ we have

$$\Gamma(X, \mathscr{O}) = R$$

so that we can recover the ring $R$ from the ringed space $(\operatorname{Spec} R, \mathscr{O})$ (at least up to isomorphism).

At the point $p \in \operatorname{Spec} R$, $\mathscr{O}$ has stalk

$$\begin{aligned} \mathscr{O}_p &= \varinjlim_{p \in D(f)} R_f = \varinjlim_{f \notin p} R_f \\ &= R_p, \end{aligned}$$

where $R_p$ is the localization $(R-p)^{-1}R$ of $R$ at the prime ideal $p$. Rings of the form $R_p$ are all local rings; that is, they each have a unique maximal ideal. In fact, as the stalk of the structure sheaf $\mathscr{O}$ at the point $p$, $R_p$ describes the nature of the ringed space Spec$R$ 'near' the point $p$, and this is the origin of the term 'local ring'.

### Definition 4.2.8

A ringed space (over $Z$) isomorphic to (Spec$R$, $\mathscr{O}$) for some ring $R$ is called an affine scheme. As the construction of (4.2.5) and (4.2.7) shows, (Spec$R$, $\mathscr{O}$) has a natural structure of the ringed space over $R$; indeed for any $S$ such that $R$ is an $S$-algebra (that is, we are given a ring morphism $S \to R$), Spec$R$ is a ringed space over $S$.

**Examples 4.2.9**

[1] Although for all fields $k$ the topological spaces Spec$k$ are homeomorphic, they are distinguished by their structure sheaves, since $k \cong \Gamma$ (Spec$k$, $\mathscr{O}$) $\cong \mathscr{O}_P$ where $P$ is the point of Spec$k$.

[2] For any ring $R$, the affine scheme Spec$R[x_1, x_2, \cdots, x_n]$ of the polynomial ring in $n$ variables is called affine $n$-space over $R$, and is denoted by $A^n_R$.

For $R = k$, an algebraically closed field, as in 4.2.4[2] and 4.2.4[3], and $S = k[x_1, \cdots, x_n]$ all the $S_f$ (for $f \in S$) can be considered as subrings of

$$k(x_1, \cdots, x_n) = \text{the field of fractions of } S.$$

At a point $(x_1 - \alpha_1, \cdots, x_n - \alpha_n)$ of $M \subseteq$ Spec$S$, $f \in S$ has the value $f(\alpha_1, \cdots, \alpha_n) \in k$; this corresponds to taking the image of $f$ in the residue field $R_p - pR_p$ of the local ring at $p = (x_1 - \alpha_1, \cdots, x_n - \alpha_n)$. Then

$$M \cap D(f) \cong \{(\alpha_1, \cdots, \alpha_n) \in k^n \mid f(\alpha_1, \cdots, \alpha_n) \neq 0\}$$

and for $s \in k(x_1, \cdots, x_n)$ and $U$ open in Spec$S = A_k^n$ we have

$$s \in \Gamma(U, \mathcal{O}) \Leftrightarrow \forall\, (\alpha_1, \cdots, \alpha_n) \in M \cap U,\ s \text{ can be written in the form } s = \frac{f}{g} \text{ with } f, g \in S \text{ and } g(\alpha_1, \cdots, \alpha_n) \neq 0.$$

Thus if we look just at $\mathcal{O} \mid M$, we see the sheaf of algebraic functions on $k^n$.

**Theorem 4.2.10**

Spec is a contravariant functor: that is, a ring morphism $r : R \to S$ gives rise in a natural way to a continuous map

$$\phi : \mathrm{Spec}\, S \to \mathrm{Spec}\, R$$

and a $\phi$-morphism of sheaves of rings $\mathcal{O}_{\mathrm{Spec}\, R} \to \mathcal{O}_{\mathrm{Spec}\, S}$. Furthermore for each $x \in \mathrm{Spec}\, S$ the stalk morphism

$$\mathcal{O}_{\mathrm{Spec}\, R,\ \phi(x)} \to \mathcal{O}_{\mathrm{Spec}\, S,\ x}$$

is a local morphism of local rings; that is, it takes the maximal ideal into the maximal ideal.

***Proof.*** If $p$ is a prime ideal of $S$, let $\phi(p) = r^{-1}[p]$; then since the ring morphism

$$R/r^{-1}[p] \to S/p$$

is injective, $\phi(p)$ is a prime ideal of $R$. Since for $f \in R$

$$\begin{aligned}\phi^{-1}[D(f)] &= \{ p \in \mathrm{Spec}\, S \mid \phi(p) \in D(f) \} \\ &= \{ p \in \mathrm{Spec}\, S \mid r^{-1}[p] \not\ni f \} = D(rf)\end{aligned}$$

the map $\phi$ is continuous. Under the map

$$R \xrightarrow{r} S \to S_{r(f)}$$

$f$ becomes a unit, so we get induced maps

$$\{\Gamma(D(f),\ \mathscr{O}) =\} R_f \to S_{rf} \{= \Gamma(D(rf),\ \mathscr{O})\}$$

giving the $\phi$-morphism required.

Similarly, since for $p \in \operatorname{Spec} S$

$$R \xrightarrow{r} S \to S_p$$

sends every $f \not\in \phi(p)$ into a unit, the induced map

$$R_{\phi(p)} \to S_p$$

of stalks sends each member of the maximal ideal $\phi(p) \cdot R_{\phi(p)}$ into the maximal ideal $pS_p$. □

## 4.3 Geometric Space

**Definition 4.3.1**

Let $R$ be a ring. A ringed space $(X,\ \mathscr{O}_X)$ over $R$ is called a geometric space (over $R$) iff all the stalks $\mathscr{O}_{X,x}$ (for $x \in X$) are local rings (that is, each has a unique maximal ideal, called $m_x$).

A morphism of geometric spaces over $R(X, \mathscr{O}_X) \to (Y, \mathscr{O}_Y)$ (or $R$-morphism) is a morphism of ringed spaces, with an underlying continuous map $\phi : X \to Y$ and $\phi$-morphism of sheaves of $R$-algebras $\psi : \mathscr{O}_Y \to \mathscr{O}_X$, with the additional property that for each $x \in X$ the stalk map

$$\psi_x : \mathscr{O}_{Y,\phi(x)} \to \mathscr{O}_{X,x}$$

is a local morphism of local rings; that is $\psi_x(m_{\phi(x)}) \subseteq m_x$.

Thus we have defined a (non-full) subcategory of the category of ringed spaces over $R$.

**Examples 4.3.2**

[1] By (4.2.7) and (4.2.8) any affine scheme Spec$R$ is a geometric space (over $R$ or $Z$), and any ring morphism $R \to S$ gives rise by (4.2.11) to a morphism of geometric spaces Spec$S \to$ Spec$R$ (again over $R$ of $Z$).

[2] Let $R$ be any ring, and $F$ an $R$-algebra which is also a topological field (so that the operations of subtraction and division are continuous); for instance $R = F = \mathbb{R}$, or $R = \mathbb{Z}$, $F = \mathbb{C}$. For any topological space $X$, let $C^F$ be the sheaf of continuous $F$-valued functions on $X$ (cf. (4.1.5[2])). Then $(X, C^F)$ is a geometric space over $R$ (see 4.3.3).

[3] The example $(X, C^r)$ of (4.1.5[3]) (with $X$ a banach space over $R$) is a geometric space over $R$ (see 4.3.3).

[4] The example $(X, C^\omega)$ of (4.1.5[4]) is a geometric space over $K$ (see 4.3.3).

### 4.3.3

In each of the cases (4.3.2[2], [3], [4]) the stalks $\mathscr{O}_{X,x}$ are local rings with maximal ideal

$$m_x = \{g \in \mathscr{O}_{X,x} \mid g \text{ is the germ of a function } f \text{ such that } f(x) = 0\}.$$

For this is clearly an ideal, and $\mathscr{O}_{X,x}/m_x \cong F$ or $R$ or $K$ (according to the case) by $g \mapsto$ value of $g$ at $x$, so that $m_x$ is maximal (since $F$, $R$, $K$ are each fields). Also $g \notin m_x$ implies that $g$ is the germ of a function $f$ such that $f(x) \neq 0$, and so in a small enough neighbourhood of $x$, $\dfrac{1}{f}$ exists; hence $g$ is a unit in $\mathscr{O}_{X,x}$, so that $m_x$ is the unique maximal ideal of $\mathscr{O}_{X,x}$.

Note that in each case the $R$-algebra structure on $\mathcal{O}_X$ is given by the maps $R \to \Gamma(U, \mathcal{O}_X)$ (for $U$ open in $X$) which send $r \in R$ to the constant function with value $r$.

### 4.3.4

When we look at geometric space morphisms between geometric spaces of the types of (4.3.2[2], [3], [4]) a remarkable thing happens. Let us examine for example the typical case of an $R$-morphism $(X, \mathcal{O}_X) \to (Y, \mathcal{O}_Y)$ where

$X = R^n$, $\mathcal{O}_X = C^r$ is the sheaf of differentiable functions on $X$;
$Y = R^m$, $\mathcal{O}_Y = C^r$ is that on $Y$.

The underlying continuous map $\phi: R^n \to R^m$ is given by say

$$\phi(x) = (\rho_1(x), \cdots, \rho_m(x)) \in R^m$$

where if $\pi_i: Y \to R$ is projection onto the $i$-th coordinate, we have

$$\rho_i = \pi_i \circ \phi.$$

Now consider the $\phi$-morphism $\psi: \mathcal{O}_Y \to \mathcal{O}_X$. This has the property that each stalk map

$$\mathcal{O}_{Y, \phi(x)} \to \mathcal{O}_{X, x} \text{ for } x \in X$$

is local; hence by the above interpretation of the maximal ideals, if $f$ is a differentiable function on some open set $V$ in $Y$, we have

$$\forall x \in \phi^{-1}(V), \quad f(\phi(x)) = 0 \Rightarrow \psi(f)(x) = 0$$

$$(\text{i.e., } f_{\phi(x)} \in m_{\phi(x)}) \quad (\text{i.e., } \psi_x(f_{\phi(x)}) \in m_x).$$

Applying this to the differentiable function $(f - c)$ with $c \in R$ we get

$$\forall x \in \phi^{-1}(V), \quad f(\phi(x)) = c \Rightarrow \psi(f)(x) = c,$$

in other words $\psi(f)(x) = f(\phi(x))$, so that

$$\psi(f) = f \circ \phi.$$

Thus the sheaf morphism $\psi$ is just 'compose with $\phi$':

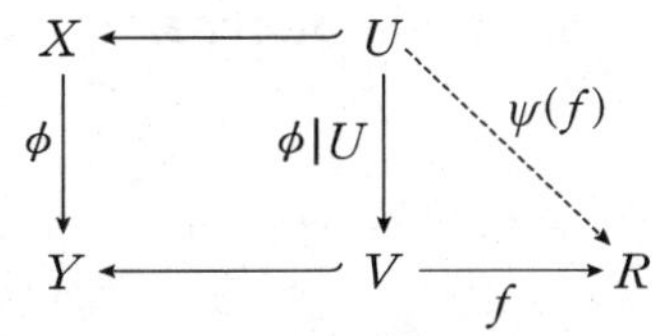


In particular, $\psi(\pi_i) = \pi_i \circ \phi = \rho_i$; but each $\pi_i$ is differentiable and so each $\psi(\pi_i) = \rho_i$ is differentiable, so that $\phi$ is actually an ($r$ times continuously) differentiable map : $R^n \to R^m$.

Thus the effect of insisting that the a prior merely continuous $\phi$ has an associated $\phi$-morphism $C_Y^r \to C_X^r$ is to ensure that $\phi$ is actually differentiable. Conversely, a differentiable map $\phi : X \to Y$ clearly defines a $\phi$-morphism $C_Y^r \to C_X^r$ by the above procedure of composing with $\phi$, and hence a geometric space morphism: $X \to Y$.

A similar argument shows that any $K$-morphism of geometric spaces $(X, C^\omega) \to (Y, C^\omega)$ (spaces as in (4.3.2[4]) has an underlying map which is analytic.

## Definition 4.3.5

Fix a ring $R$. Suppose we have a geometric space $(M, \mathscr{O}_M)$ over $R$ (to be regarded as a 'model'). We say that a geometric space $(X, \mathscr{O}_X)$ over $R$ is locally isomorphic to $(M, \mathscr{O}_M)$ iff for any $x \in X$ there is an open neighbourhood $U$ of $x$ in $X$ and an open set $V$ in $M$ such that there is an isomorphism

$$(U, \mathscr{O}_X \mid U) \cong (V, \mathscr{O}_M \mid V)$$

as geometric spaces over $R$.

Given a class $\mathfrak{W}$ of model spaces $(M, \mathscr{O}_M)$, we say that $(X, \mathscr{O}_X)$ is a manifold of type $\mathfrak{W}$ (over $R$) iff $X$ can be covered by open sets $U$ such that each $\mathscr{O}_X \mid U$ is locally isomorphic to some $(M, \mathscr{O}_M) \in \mathfrak{W}$.

A morphism between manifolds of type $\mathfrak{W}$ is just a morphism of geometric spaces over $R$.

Some authors use the word variety in place of manifold.

**Examples 4.3.6**

[1] Let $R = \mathbb{R}$.

Putting $\mathfrak{W} = \{(\mathbb{R}^n, C^R)\}$ we obtain topological manifolds of dimension $n$.

Putting $\mathfrak{W} = \{(\mathbb{R}^n, C^r)\}$ we obtain differentiable manifolds of class $r$ and dimension $n$.

Putting $\mathfrak{W} = \{(\mathbb{R}^n, C^\omega) \mid n \in \mathbb{N}\}$ we obtain real analytic manifolds.

[2] Let $R = \mathbb{C}$.

Putting $\mathfrak{W} = \{(\mathbb{C}, C^\omega)\}$ we obtain Riemann surfaces.

Putting $\mathfrak{W} = \{(\mathbb{C}^n, C^\omega) \mid n \in \mathbb{N}\}$ we obtain complex analytic manifolds.

[3] Let $R = \mathbb{Q}_p$.

Putting $\mathfrak{W} = \{(\mathbb{Q}_p^n, C^\omega) \mid n \in \mathbb{N}\}$ we obtain $p$-adic analytic manifolds.

Putting $\mathfrak{W} = \{(E, C^\omega) \mid E \text{ is a hilbert space over } \mathbb{Q}_p\}$ we obtain $p$-adic analytic manifolds of Hilbert type.

[4] Let $R = \mathbb{Z}$

Putting $\mathfrak{W} =$ the class of all affine schemes, we obtain schemes.

**Theorem 4.3.7**

Let $(X, \mathscr{O}_X)$ be a geometric space over $\mathbb{Z}$, and let $R$ be a ring. For each ring morphism $t : R \to \Gamma(X, \mathscr{O}_X)$ there is a unique morphism of geometric spaces

$$G: (X, \mathscr{O}_X) \to (\mathrm{Spec}\, R, \mathscr{O})$$

such that the induced map

$$R \xrightarrow{\sim} \Gamma(\mathrm{Spec}\, R, \mathscr{O}) \to \Gamma(X, \mathscr{O}_X)$$

is $t$. In other words the map

$$\mathrm{Hom}_{\mathrm{Gsp}}((X, \mathscr{O}_X), (\mathrm{Spec}\, R, \mathscr{O})) \to \mathrm{Hom}_{\mathrm{Ring}}(\mathrm{R}, \Gamma(X, \mathscr{O}_X))$$
$$: \Phi \mapsto `\Gamma(X, \Phi)'$$

is bijective (where Gsp is the category of geometric spaces over $\mathbb{Z}$).

***Proof.*** Given $t$, we first construct the underlying continuous map $g$ of $G$. For $x \in X$, let $t_x$ be defined to make the diagram

$$\begin{array}{ccc} & & \mathscr{O}_{X,x} \\ & \overset{t_x}{\nearrow} & \uparrow \\ R & \xrightarrow[t]{} & \Gamma(X, \mathscr{O}_X) \end{array}$$

commute. Let $g(x) = t_x^{-1}(m_x)$ where $m_x$ is the maximal ideal of $\mathscr{O}_{X,x}$; then $g(x) \in \mathrm{Spec}\, R$. For $f \in R$ we have

$$\begin{aligned} g^{-1}(D(f)) &= \{x \in X \mid g(x) \not\ni f\} \\ &= \{x \in X \mid t_x(f) \notin m_x\} \end{aligned}$$

and this is open in $X$, for $f' = t_x(f) \notin m_x$ holds iff $f'$ is a unit in $\mathscr{O}_{X,x}$, so there is $g' \in \mathscr{O}_{X,x}$ such that $f' \cdot g' = 1$, and this equality still holds in some neighbourhood of $x$.

This calculation also shows that the map

$$R \to \Gamma(X, \mathscr{O}_X) \to \Gamma(g^{-1}(D(f)), \mathscr{O}_X)$$

inverts $f$, so it factorizes as

$$\Gamma(D(f), \mathcal{O}) = R_f \to \Gamma(g^{-1}(D(f)), \mathcal{O}_X)$$

and this defines the morphism $G$, which clearly has $t$ as its associated global section map.

The uniqueness follows from the easily verified fact that this construction is inverse to the process of taking global sections (so that the map given in the Theorem is bijective). □

## 4.4 Module over Ringed Space

### *4.4.1*

Recall that if $R$ is a ring with a one, a (left) $R$-module is an additive abelian group $M$ equipped with a map

$$R \times M \to M: (r, m) \mapsto rm$$

(that is, an operation of $R$ on $M$) satisfying the conditions:

(i) $\forall r \in R \quad \forall m, n \in M \qquad r(m+n) = rm + rn$

(ii) $\forall r, s \in R \quad \forall m \in M \qquad (r+s)m = rm + sm$ and $(rs)m = r(sm)$

(iii) $\forall m \in M \qquad 1m = m.$

A morphism $f: M \to N$ of $R$-modules is an abelian group morphism which satisfies the additional condition:

$$\forall m \in M \quad \forall r \in R \qquad f(rm) = rf(m)$$

(so that $f$ respects (or preserves) the operation of $R$).

The category $R$-mod of all $R$-modules is abelian (even if $R$ is not commutative); indeed, the kernels, cokernels and biproducts are constructed as if in the category Abgp, and are then seen to carry a natural

$R$-module structure.

### *4.4.2*

More generally, if $\rho$: $R \to S$ is a (one-preserving) ring morphism, and $M$ is an $R$-module and $N$ is an $S$-module, then a map $f : M \to N$ is called a morphism over $\rho$ (or dihomomorphism) iff $f$ is an abelian group morphism and

$$\forall r \in R \;\; \forall m \in M, \qquad f(rm) = \rho(r) f(m).$$

**Example 4.4.3**

If $R'$ is an $R$-algebra, with structure map $\alpha : R \to R'$, then $R'$ is an $R$-module with the operation of $R$ defined by

$$rr' = \alpha(r) \cdot r' \;\; \text{(product in } R'\text{)} \quad \text{for } r \in R,\ r' \in R'.$$

A commutative square of ring morphisms:

$$\begin{array}{ccc} R' & \xrightarrow{\rho'} & S' \\ {\scriptstyle \alpha}\uparrow & & \uparrow{\scriptstyle \beta} \\ R & \xrightarrow{\rho} & S \end{array}$$

may be regarded as giving a module morphism $\rho' : R' \to S'$ over $\rho$.

## Definition 4.4.4

Suppose that $(X, \mathscr{O})$ is a ringed space over a ring $R$. An $\mathscr{O}$-Module (note the use of the capital letter), or sheaf of $\mathscr{O}$-modules, $M$ is a sheaf of abelian groups on $X$ with the additional structure that

(i) for each open $U$ in $X$, $\Gamma(U, M)$ is a $\Gamma(U, \mathscr{O})$-module; and

(ii) wherever $V \subseteq U$ are open in $X$ the restriction map

$$\Gamma(U, M) \to \Gamma(V, M)$$

is a module morphism over the ring morphism (restriction)

$$\Gamma(U, \mathscr{O}) \to \Gamma(V, \mathscr{O}).$$

There is an obvious definition of morphism for $\mathscr{O}$-Modules: $f : M \to N$ must be a morphism of abelian sheaves, and for each open $U$ in $X$

$$\Gamma(U, f) : \Gamma(U, M) \to \Gamma(U, N)$$

must be a $\Gamma(U, \mathscr{O})$-module morphism.

Hence we have defined a category $\mathscr{O}$-Mod of all $\mathscr{O}$-Modules on $X$.

**Examples 4.4.5**

[1] For any topological space $X$, let $\mathscr{O}$ be the constant sheaf $\mathbb{Z}$ on $X$ (as in (4.1.5[2])). Then any abelian sheaf on $X$ is an $\mathscr{O}$-Module, and the category $\mathscr{O}$-Mod is just the category Shv/$X$.

[2] Let $R$ be a ring, $X = \operatorname{Spec} R$ and $\mathscr{O}$ be the structure sheaf on $X$ as in (4.2.7). Starting with an $R$-module $M$, we can construct an $\mathscr{O}$-Module $\widetilde{M}$ on $\operatorname{Spec} R$ as follows.

To each basic open set $D(f)$ of $\operatorname{Spec} R$ we associate the $R_f$-module $M_f = S^{-1}M$ where $S = \{f^n \mid n \in \mathbb{N}\}$. The restriction maps follow from Commutative Algebra, and the sequence

$$M_f \to \prod_{\lambda \in \Lambda} M_{f_\lambda} \rightrightarrows \prod_{(\lambda, \mu) \in \Lambda \times \Lambda} M_{f_\lambda f_\mu}$$

corresponding to an open cover $D(f) = \bigcup_{\lambda \in \Lambda} D(f_\lambda)$ is an equaliser.

Hence by Lemma 4.2.6, $M$ defines a sheaf $\widetilde{M}$ on $\operatorname{Spec} R$ which is easily seen to be an $\mathscr{O}$-Module.

We remark in passing that $\widetilde{R} \cong \mathscr{O}$, and that it can be shown that the category $R$-mod is equivalent to a full subcategory of $\mathcal{O}$-Mod by the assignment $M \mapsto \widetilde{M}$. An $\mathcal{O}_X$-Module on a scheme $(X, \mathcal{O}_X)$ which is given locally (on an open cover by affine schemes) by Modules of the form $\widetilde{M}$ is called quasi-coherent.

[3] Suppose that $(X, \mathcal{O}_X)$ is a differentiable or analytic manifold in the interpretation of (4.3.6). Then for $p \in N$ the assignment

$$U \mapsto \text{the set of } p\text{-forms on } U \text{ (for } U \text{ open in } X)$$

defines an $\mathcal{O}_X$-Module.

Similarly the sheaves of sections (4.2.3[3]) of vector bundles over a topological, differentiable or analytic manifold $(X, \mathcal{O}_X)$ can be regarded as $\mathcal{O}_X$-Modules.

### Construction 4.4.6

Recall that if $R$ is a ring and $(M_\lambda)_{\lambda \in \Lambda}$ a family of $R$-modules, then the direct product $\prod_{\lambda \in \Lambda} M_\lambda$ is the $R$-module defined by giving the product set pointwise operations. The direct sum $\bigoplus_{\lambda \in \Lambda} M_\lambda$ is the sub-$R$-module of $\prod_{\lambda \in \Lambda} M_\lambda$ generated by the images of all the injection maps

$$i_\mu : M_\mu \to \prod_{\lambda \in \Lambda} M_\lambda : m \mapsto (M_\lambda)_{\lambda \in \Lambda}$$

where $m_\lambda = \begin{cases} m & \text{if } \lambda = \mu \\ 0 & \text{if } \lambda \neq \mu \end{cases}$ so that $(m_\lambda)_{\lambda \in \Lambda} \in \prod_{\lambda \in \Lambda} M_\lambda$ is an element of the direct sum iff for all but finitely many of the $\lambda \in \Lambda$, $m_\lambda = 0$. Each of these constructions satisfies a universal property. For any $R$-module $N$ there are bijections:

$$(*) \begin{cases} \operatorname{Hom}(N, \prod_{\lambda \in \Lambda} M_\lambda) \xrightarrow{\sim} \prod_{\lambda \in \Lambda} \operatorname{Hom}(N, M_\lambda) : f \mapsto (p_\lambda \circ f)_{\lambda \in \Lambda} \\ \operatorname{Hom}(\bigoplus_{\lambda \in \Lambda} M_\lambda, N) \xrightarrow{\sim} \prod_{\lambda \in \Lambda} \operatorname{Hom}(M_\lambda, N) : f \mapsto (f \circ i_\lambda)_{\lambda \in \Lambda} \end{cases}$$

where $p_\mu : \prod_{\lambda \in \Lambda} M_\lambda \to M_\mu$ is the $\mu$-th projection.

If now $(X, \mathscr{O})$ is a ringed space over $R$ and $(M_\lambda)_{\lambda \in \Lambda}$ is a family of $\mathscr{O}$-Modules, it is easily verified that the following presheaves are sheaves:

$$\left.\begin{aligned} U &\mapsto \prod_{\lambda \in \Lambda} \Gamma(U, M_\lambda) \\ U &\mapsto \bigoplus_{\lambda \in \Lambda} \Gamma(U, M_\lambda) \end{aligned}\right\} \text{ for } U \text{ open in } X$$

and so define $\mathscr{O}$-Modules called the direct product $\prod_{\lambda \in \Lambda} M_\lambda$ and the direct sum $\bigoplus_{\lambda \in \Lambda} M_\lambda$ of the family. There are natural projection morphisms $p_\mu$ : $\prod_{\lambda \in \Lambda} M_\lambda \to M_\mu$ and injection morphisms $i_\mu : M_\mu \to \bigoplus_{\lambda \in \Lambda} M_\lambda$, and the universal properties ( $*$ ) hold as stated, in the category $\mathscr{O}$-Mod.

Since direct sums and stalks are both defined as colimits, we have for $x \in X$ the stalk:

$$\begin{aligned} (\bigoplus_{\lambda \in \Lambda} M_\lambda)_X &= \varinjlim_{x \in U} (\bigoplus_{\lambda \in \Lambda} \Gamma(U, M_\lambda)) \\ &= \bigoplus_{\lambda \in \Lambda} \varinjlim_{x \in U} \Gamma(U, M_\lambda) \\ &= \bigoplus_{\lambda \in \Lambda} (M_\lambda)_X \end{aligned}$$

as can easily be verified directly. However there is no corresponding expression for the stalks of the direct product.

### Construction 4.4.7

Let $(X, \mathscr{O})$ be a ringed space over a ring $R$. If $L$, $M$ are $\mathscr{O}$-Modules we can define a presheaf on $X$ by

$$U \to \Gamma(U, \Gamma) \otimes_{\Gamma(U, \mathscr{O})} \Gamma(U, M)$$

for $U$ open in $X$ (tensor product of $\Gamma(U, \mathscr{O})$-modules). The sheafification of this presheaf is an $\mathscr{O}$-Module called the tensor product of $L$ and $M$ over $\mathscr{O}$ and denoted by $L \otimes_{\mathscr{O}} M$.

Since sheafification does not change the stalks, and tensor product commutes with direct limits, $L \otimes_{\mathscr{O}} M$ has the stalk

$$(L \otimes_{\mathscr{O}} M)_X = \varinjlim_{x \in U} \Gamma(U, L) \otimes_{\Gamma(U, \mathscr{O})} \Gamma(U, M) = L_X \otimes_{\mathscr{O}_x} M_X$$

at $x \in X$; also if $s, t$ are sections of $L$, $M$ respectively over an open set $U$ then the map

$$x \mapsto s_x \otimes t_x \text{ for } x \in U$$

is a section of $L \otimes_{\mathscr{O}} M$ over $U$.

It is easy to see from the definition that for any $\mathscr{O}$-Module $M$ we have $M \otimes_{\mathscr{O}} \mathscr{O} \cong M$.

#### *4.4.8*

Tensor product of Modules inherits many of the properties of tensor product of modules; for instance, tensor product with a fixed $\mathscr{O}$-Module $M$ gives a right exact covariant functor

$$M \otimes_{\mathscr{O}} - : \mathscr{O}\text{-Mod} \to \mathscr{O}\text{-Mod}.$$

There is also a universal property as follows. If $L$, $M$, $N$ are $\mathscr{O}$-Modules, a bilinear map from $L$, $M$ to $N$ is a map of sheaves of sets $f : L \oplus M \to N$ such that for each open set $U$ the map

$$\Gamma(U, f) : \Gamma(U, L) \oplus \Gamma(U, M) \to \Gamma(U, N)$$

is a bilinear map of $\Gamma(U, \mathscr{O})$-modules. Such maps form a set Bilin($L$, $M$; $N$) which is covariant-functorial in $N$.

**Theorem 4.4.9**

$L \otimes_{\mathscr{O}} M$ represents the functor Bilin($L$, $M$; -) : $\mathscr{O}$-Mod→Sets; that is, there is an isomorphism

$$\mathrm{Bilin}(L, M; N) \xrightarrow{\sim} \mathrm{Hom}_{\mathscr{O}}(L \otimes_{\mathscr{O}} M, N)$$

which is natural in the $\mathscr{O}$-Module $N$. This property characterizes the $\mathscr{O}$-Module $L \otimes_{\mathscr{O}} M$ up to isomorphism.

***Proof.*** Direct from the corresponding proposition for modules over a ring, and the universal property of sheafification. □

Suppose that $(X, \mathscr{O})$ is a ringed space over a ring $R$, and $U$ is open in $X$. Then $U$ is a ringed space with structure sheaf $\mathscr{O} \mid U$, and if $M$ is an $\mathscr{O}$-Module, it is easy to see that $M \mid U$ is in a natural way an $(\mathscr{O} \mid U)$-Module.

**Theorem 4.4.10**

If $L$, $M$ are $\mathscr{O}$-Modules and $U$ is open in $X$, then there is a natural isomorphism of $(\mathscr{O} \mid U)$-Modules

$$(L \mid U) \otimes \mathscr{O} \mid U (M \mid U) \xrightarrow{\sim} (L \otimes_{\mathscr{O}} M) \mid U.$$

***Proof.*** There is a natural bilinear map

$$(L \mid U) \oplus (M \mid U) \to (L \otimes_{\mathscr{O}} M) \mid U$$

obtained by restricting $L \oplus M \to L \otimes_{\mathscr{O}} M$ to $U$. By (4.4.8) this induces the required map, which is an isomorphism since on the stalk at $x \in U$ it reduces, after the identification of (4.4.7), to the identity map

$$L_x \otimes M_x \to L_x \otimes M_x.$$

□

**Construction 4.4.11**

Let $(X, \mathscr{O})$ be a ringed space over a ring $R$. If $M$, $N$ are $\mathscr{O}$-Modules, define the presheaf $\mathrm{Hom}(M, N)$ by

$$\mathrm{Hom}(M, N)(U) = \mathrm{Hom}_{\mathscr{O}|U}(M|U, N|U) \quad \text{for } U \text{ open in } X,$$

where the right-hand side is the set of $(\mathscr{O}|U)$-Module morphisms of $M|U$ into $N|U$, and has a natural structure of $\Gamma(U, \mathscr{O})$-module. If $U \supseteq V$ are open, we have a map

$$\mathrm{Hom}_{\mathscr{O}|U}(M|U, N|U) \xrightarrow{\alpha^*} \mathrm{Hom}_{\mathscr{O}|V}(M|V, N|V)$$

where $\alpha$: $V \hookrightarrow U$ is the inclusion, and $\alpha^*$ is easily seen to be a module morphism over the restriction $\Gamma(U, \mathscr{O}) \to \Gamma(V, \mathscr{O})$. If we reinterpret the elements of $\mathrm{Hom}_{\mathscr{O}|U}(M|U, N|U)$ as continuous maps between certain subspaces of the sheaf spaces $LM$ and $LN$, we can check readily that $\mathrm{Hom}(M, N)$ is a sheaf of $\mathscr{O}$-modules, called the sheaf of germs of homomorphisms from $M$ to $N$, or sheaf-hom for short; if we wish to emphasize the structure sheaf we can write $\mathrm{Hom}_{\mathscr{O}}(M, N)$.

***4.4.12***

Suppose that $\Phi : (X, \mathscr{O}_X) \to (Y, \mathscr{O}_Y)$ is a morphism of ringed spaces over a ring $R$, with underlying continuous map $\phi : X \to Y$ and morphism of sheaves of $R$-algebras $\psi : \mathscr{O}_Y \to \phi_* \mathscr{O}_X$.

If $M$ is an $\mathscr{O}_X$-Module, the definition
$\Gamma(V, \phi_* M) = \Gamma(\phi^{-1} V, M)$ for $V$ open in $Y$

shows that $\phi_* M$ is naturally a $(\phi_* \mathscr{O}_X)$-Module. The morphism $\psi$ gives for each open $V$ in $Y$ a change-of-rings morphism

$$\Gamma(V, \mathscr{O}_Y) \to \Gamma(V, \phi_* \mathscr{O}_X)$$

which enables us to regard $\phi_* M$ as an $\mathscr{O}_Y$-Module. Hence $\Phi$ induces a functor

$$\Phi_* : \mathscr{O}_X\text{-Mod} \to \mathscr{O}_Y\text{-Mod}.$$

If $N$ is an $\mathscr{O}_Y$-Module, then $\phi^* N$ is a sheaf of abelian groups on $X$, and an easy examination of its groups of sections shows that $\phi^* N$ has a natural structure of $(\phi^* \mathscr{O}_Y)$-Module. We also have a change-of-rings morphism $\phi^* \mathscr{O}_Y \to \mathscr{O}_X$ deduced from $\psi$, and we define

$$\Phi^* N = \phi^* N \otimes_{\phi^* \mathscr{O}_Y} \mathscr{O}_X$$

to get a functor

$$\Phi^* : \mathscr{O}_Y\text{-Mod} \to \mathscr{O}_X\text{-Mod}.$$

**Theorem 4.4.13**

If $\Phi : (X, \mathscr{O}_X) \to (Y, \mathscr{O}_Y)$ is a morphism of ringed spaces over a ring $R$, then $\Phi^*$ is left adjoint to $\Phi_*$; that is there is a natural bijection

$$\mathrm{Hom}_{\mathscr{O}_X}(\Phi^* N, M) \xrightarrow{\sim} \mathrm{Hom}_{\mathscr{O}_Y}(N, \Phi_* M)$$

whenever $M \in Ob\mathscr{O}_X$-Mod and $N \in Ob\mathscr{O}_Y$-Mod.

***Proof.*** Let $\phi : X \to Y$ be the underlying continuous map of $\Phi$.

Then we can construct the following diagram:

$$\begin{array}{ccc}
\mathrm{Hom}_{Z_X}(\phi^* N, M) & \xrightarrow[f]{\sim} & \mathrm{Hom}_{Z_Y}(N, \phi_* M) \\
\uparrow & & \uparrow \\
\mathrm{Hom}_{\phi^* \mathscr{O}_Y}(\phi^* N, M) & \xrightarrow[g]{\sim} & \mathrm{Hom}_{\mathscr{O}_Y}(N, \phi_* M) \\
\wr\uparrow h & & \\
\mathrm{Hom}_{\mathscr{O}_X}(\phi^* N \otimes_{\phi^* \mathscr{O}_Y} \mathscr{O}_X, M) & &
\end{array}$$

where $f$ is bijective, and is easily seen to induce the bijection $g$, while the map $h$ is given by composition with the morphism

$$\phi^* N \to \phi^* N \otimes_{\phi^* \mathscr{O}_Y} \mathscr{O}_X$$

(given on sections by $s \mapsto s \otimes 1$) and has inverse sending a morphism $u : \phi^* N \to M$ to the composite

$$\phi^* N \otimes_{\phi^* \mathscr{O}_Y} \mathscr{O}_X \xrightarrow{u \otimes id} M \otimes_{\phi^* \mathscr{O}_Y} \mathscr{O}_X \xrightarrow{v} M,$$

where the multiplication morphism $v$ is given on sections by $s \otimes t \mapsto ts$ (since $M$ is an $\mathscr{O}_X$-Module). □

## 4.5 Locally Free Module

### Definition 4.5.1

If $(X, \mathscr{O})$ is a ringed space over a ring $R$, an $\mathscr{O}$-Module $M$ is called locally free of rank $n \in N$ iff $X$ can be covered by open sets $U$ such that

$$M | U \cong (\mathscr{O} | U)^n \quad \text{as} \quad (\mathscr{O} | U)\text{-Modules}$$

where for a Module $N$, $N^n = \oplus_{i=1}^n N$ is the biproduct of $n$ copies of $N$.

### *4.5.2*

If $(X, \mathscr{O})$ is a ringed space over a ring $R$, the endomorphism ring of $\mathscr{O}$

$$\text{End}(\mathscr{O}) = \text{Hom}_{\mathscr{O}}(\mathscr{O}, \mathscr{O})$$

is the set of $\mathscr{O}$-Module morphisms : $\mathscr{O} \to \mathscr{O}$, with composition as multiplication. Its group of units is the group Aut($\mathscr{O}$) of $\mathscr{O}$-Module automorphisms of $\mathscr{O}$. We have a map

$$\Gamma(X, \mathscr{O}) \to \mathrm{End}(\mathscr{O})$$

sending $s \in \Gamma(M, \mathscr{O})$ to the endomorphism which is given over an open $U$ by multiplication by the restriction $\rho_U^X(s)$:

$$\Gamma(U, \mathscr{O}) \to \Gamma(U, \mathscr{O}) : t \mapsto \rho_U^X(s) \cdot t.$$

**Theorem 4.5.3**

In the situation of (4.5.2), the map

$$\Gamma(M, \mathscr{O}) \to \mathrm{End}(\mathscr{O})$$

is a ring isomorphism. Hence $\mathrm{Aut}(\mathscr{O}) \xrightarrow{\sim} \Gamma(X, \mathscr{O})^*$ (the group of units of the ring $\Gamma(X, \mathscr{O})$) and for $n \in N$

$$\mathrm{End}(\mathscr{O}^n) \xrightarrow{\sim} M_n(\Gamma(X, \mathscr{O})) = \mathrm{End}_{\Gamma(X,\mathscr{O})}(\Gamma(X, \mathscr{O})^n)$$

can be identified with the ring of $n \times n$ matrices over $\Gamma(X, \mathscr{O})$, while

$$\mathrm{Aut}(\mathscr{O}^n) \xrightarrow{\sim} GL_n(\Gamma(X, \mathscr{O})) = \mathrm{Aut}_{\Gamma(X,\mathscr{O})}(\Gamma(X, \mathscr{O})^n).$$

***Proof.*** The map of (4.5.2) is clearly a ring morphism, and it is injective, since if $s \in \Gamma(X, \mathscr{O})$ gives the zero morphism $\mathscr{O} \to \mathscr{O}$, then in particular

$$s = \rho_X^X(s) \cdot 1 = 0.$$

Given $f \in \mathrm{End}(\mathscr{O})$, each $\Gamma(U, f)$: $\Gamma(U, \mathscr{O}) \to \Gamma(U, \mathscr{O})$ is a $\Gamma(U, \mathscr{O})$-module morphism, and so is multiplication by some $t_U \in \Gamma(U, \mathscr{O})$ (namely $t_U = \Gamma(U, f)(1)$); then the commutativity of the square

$$\begin{array}{ccc} \Gamma(X, \mathscr{O}) & \longrightarrow & \Gamma(X, \mathscr{O}) \\ \downarrow & & \downarrow \\ \Gamma(U, \mathscr{O}) & \longrightarrow & \Gamma(U, \mathscr{O}) \end{array}$$

shows that $t_U = \rho_U^X(t_X)$. Hence $f \mapsto t_X = \Gamma(U, f)(1)$ gives an inverse to the morphism of (4.5.2).

If multiplication by $s \in \Gamma(X, \mathcal{O})$ is an automorphism of $\mathcal{O}$, then the map $(t \mapsto s \cdot t)$ $t$ is surjective on $\Gamma(X, \mathcal{O})$, so $s$ is a unit; the converse is clear.

The last result follows from the usual procedure of considering the various composites

$$\mathcal{O} \to \mathcal{O}^n \to \mathcal{O}^n \to \mathcal{O}$$

to obtain the matrix components. □

**Example 4.5.4**

If $X$ is a manifold of the kind described in (4.3.6[1], [2], [3]) with structure sheaf $\mathcal{O}$, then the locally free $\mathcal{O}$-Modules of rank $n$ are just the sheaves of sections of rank $n$ vector bundles. For it is easy to see that such a sheaf is locally free, by considering its restriction to the open sets of a trivialization.

Conversely, given a locally free Module $M$, we can patch together copies of $R^n$ over the intersection $U \cap V$ of two open sets over which $M$ is free by means of the isomorphism

$$\begin{aligned}(\mathcal{O} \mid U \cap V)^n &\cong (\mathcal{O} \mid U)^n \mid V \cong M \mid U \cap V \\ &\cong (\mathcal{O} \mid V)^n \mid U \cong (\mathcal{O} \mid U \cap V)^n\end{aligned}$$

which by (4.5.3) is given by a member of the general linear group

$$GL_n(\Gamma(U \cap V, \mathcal{O}))$$

(that is, an invertible $n \times n$ matrix of $R$-valued functions on $U \cap V$). □

TER

# 5 Sheaf Cohomology

Let $X$ be a topological space and $0 \to S' \to S \to S'' \to 0$ an **exact sequence of sheaves** on $X$, i.e., for each $x \in X$ the sequence of stalks of abelian groups

$$0 \to S'(x) \to S(x) \to S''(x) \to 0$$

is exact.

Then we have an exact sequence of abelian groups of global sections

$$0 \to \Gamma(X, S') \to \Gamma(X, S) \to \Gamma(X, S''),$$

but in general $\Gamma(X, S) \to \Gamma(X, S'')$ is not an epimorphism.

**Example**

Let $X = \mathbb{CP}^1$, $S$ the sheaf of holomorphic functions on $X$, and $S''$ the sheaf concentrated on $n$ points of $X$ (see Chapter 3, Example [7]), where $n > 1$. Let $\varphi : S \to S''$ be the homomorphism defined in Chapter 3, Example [9].

Then the $\varphi$ is a sheaf epimorphism, but the induced homomorphism is not an epimorphism since $\Gamma(X, S) = \mathbb{C}$ by **Liouville Theorem** : a global holomorphic function on $X$ is constant and

$$\Gamma(X, S'') \simeq \mathbb{C}^n,$$

and the map

$$\varphi(X) : \Gamma(X, S) \to \Gamma(X, S'')$$

is the diagonal map $\mathbb{C} \to \mathbb{C}^n$.

**[Axiomatic Sheaf Cohomolgy]**

We want to measure the amount by which $\Gamma(X, S) \to \Gamma(X, S'')$ fails to be an epimorphism, i.e., how many sections of $S''$ do not come from sections of $S$.

Our goal is to define cohomolgy groups $H^q(X, S)$ of $X$ with coefficients in sheaf $S$ for each integer $q \geq 0$ which are abelian groups and satisfy the following axioms:

($C_1$) Each $H^q(X, S)$ is a covariant functor in $S$,

i.e., a sheaf homomorphism $S \to T$ induces cohomology homomorphisms $H^q(X, S) \to H^q(X, T)$.

($C_2$) $H^0(X, S) = \Gamma(X, S)$.

($C_3$) If $0 \to S' \to S \to S'' \to 0$ is an exact sequence of sheaves on $X$ then there is a connecting homomorphism

$$H^q(X, S'') \to H^{q+1}(X, S')$$

for each $q$ such that we have a cohomology long exact sequence

$$0 \to H^0(X, S') \to H^0(X, S) \to H^0(X, S'') \to H^1(X, S') \to H^1(X, S) \to \cdots.$$

Also the sheaf cohomology satisfies the following functorialities :

(1) The identity homomorphism $id : S \to S$ induces the identity homomorphism $id : H^q(X, S) \to H^q(X, S)$.

(2) If the diagram

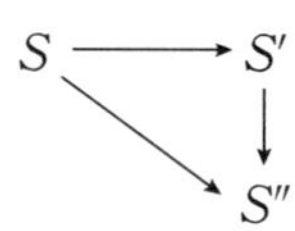


commutes, then for each $q$ the induced diagram

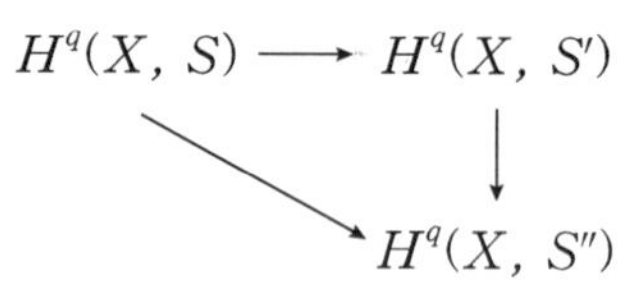


is also commutative.

(3) For every homomorphism of short exact sequences of sheaves

$$\begin{array}{ccccccccc} 0 & \longrightarrow & S & \longrightarrow & S' & \longrightarrow & S'' & \longrightarrow & 0 \\ & & \downarrow & & \downarrow & & \downarrow & & \\ 0 & \longrightarrow & T & \longrightarrow & T' & \longrightarrow & T'' & \longrightarrow & 0 \end{array},$$

the induced diagram of connecting homomorphisms

$$\begin{array}{ccc} H^q(X, S'') & \longrightarrow & H^{q+1}(X, S) \\ \downarrow & & \downarrow \\ H^q(X, T'') & \longrightarrow & H^{q+1}(X, T) \end{array}$$

commutes.

We shall give two constructions.

The first is a generalization of Čech cohomology but does not satisfy $(C_3)$ unless restrictions are placed on the base space $X$.

The second is the Grothendieck cohomology which depends on homological algebra and gives a good theory for any space $X$ which may have bad its topology.

## 5.1 Čech Cohomology

**Definition 5.1.1**

Let $X$ be a topological space, $F$ a presheaf on $X$, $\mathscr{U} = \{U_i | i \in I\}$ an open covering of $X$. A $q$**-cochain** of the covering $\mathscr{U}$ with values in $F$ is defined to be a family $f = \{f_{i_0 \cdots i_q}\}$,

$$f_{i_0 \cdots i_q} \in F(U_{i_0} \cap \cdots \cap U_{i_q})$$

such that $f_{i_0 \cdots i_q}$ is an alternating function of the indices $i_0, \cdots, i_q$. The set of all $q$-cochains forms an abelian group denoted by

$$C^q(\mathscr{U}, F).$$

We define the **coboundary** homomorphism

$$\delta^q : C^q(\mathscr{U}, F) \to C^{q+1}(\mathscr{U}, F), \; q \geq 0$$

as follows:

$$\delta^q(f)_{i_0 \cdots i_{q+1}} = \sum_{j=0}^{q+1} (-1)^j f_{i_0 \cdots \hat{i_j} \cdots i_{q+1}}, \quad \delta^{-1} = 0$$

where the symbol underneath $\wedge$ means that it is to be omitted and $f_{i_0 \cdots \hat{i_j} \cdots i_{q+1}} \in F(U_{i_0} \cap \cdots \cap \widehat{U_{i_j}} \cap \cdots \cap U_{i_{q+1}})$.

**Theorem 5.1.2**

$$\delta^{q+1} \circ \delta^q = 0$$

## Definition 5.1.3

We define the **cohomology groups of $\mathscr{U}$ with coefficients in $F$** by

$$H^q(\mathscr{U},\ F) = \mathrm{Ker}\ \delta^q / \mathrm{Im}\ \delta^{q-1}.$$

Let $\mathscr{U} = \{V_j | j \in J\}$ be another open covering of $X$ and let $\mathscr{V}$ **refine** $\mathscr{U}$ i.e., each open set in $\mathscr{V}$ is contained in some open sets of $\mathscr{U}$. There is a map $\tau: J \to I$ such that $V_j \subseteq U_{\tau(j)}$ for each $j \in J$. If $f \in C^q(\mathscr{U},\ F)$, we define $\tau f \in C^q(\mathscr{V},\ F)$ by the restriction,

$$(\tau f)_{j_0 \cdots j_q} = \rho^{U_{\tau(j_0)} \cap \cdots \cap U_{\tau(j_q)}}_{V_{j_0} \cap \cdots \cap V_{j_q}} (f_{\tau(j_0) \cdots \tau(j_q)}) \in F(V_{j_0} \cap \cdots \cap V_{j_q})$$

Thus we have a homomorphism

$$\tau : C^q(\mathscr{U},\ F) \to C^q(\mathscr{V},\ F)$$

for each integer $q \geq 0$.

The $\tau$ commutes with $\delta$ and therefore induces **cohomology homomorphisms**

$$\theta_{uv} : H^q(\mathscr{U},\ F) \to H^q(\mathscr{V},\ F), \quad q \geq 0.$$

If $\tau_1 : J \to I$ is another choice of $\tau$, then $\tau$ and $\tau_1$ induce a **cochain homotopy map**

$$H : C^*(\mathscr{U},\ F) \to C^{*-1}(\mathscr{V},\ F)$$

such that

$$\tau - \tau_1 = \delta H + H\delta.$$

Thus the homomorphism $\theta_{uv}$ is independent of the choice of $\tau$.

The set of all open coverings of $X$ forms a direct set by refinement.

(1) If $\mathscr{U}$ is an open covering of $X$ then $\theta_{uu}$ = identity.

(2) If $\mathscr{W}$ is a refinement of $\mathscr{V}$ and $\mathscr{V}$ is a refinement of $\mathscr{U}$ then

$\theta_{uw} = \theta_{vw}\theta_{uv}$.

Thus we have a direct system

$\{\theta_{uv} : H^*(\mathscr{U}, F) \to H^*(\mathscr{V}, F) \mid \mathscr{V}$ is a refinement of an open covering $\mathscr{U}$ of $X\}$

A **proper open covering** $\mathscr{U} = \{U_i\}$ of $X$ is one in which $U_i \neq U_j$ if $i \neq j$. We only consider the proper open coverings of $X$, since the class of all open coverings of $X$ is not a set.

Now we define the **$q$-th Čech cohomology group** of $X$ with coefficients in $F$ by the direct limit of the direct system of proper open coverings of $X$:

$$\check{H}^q(X, F) := \varinjlim_{\mathscr{U}} H^q(\mathscr{U}, F).$$

Note that for simplicity we will omit the check $\vee$ in Čech cohomology groups.

If $0 \to F' \to F \to F'' \to 0$ is an exact sequence of presheaves on $X$, then for each open $U \subset X$, we have exact sequences of abelian groups

$$0 \to F'(U) \to F(U) \to F''(U) \to 0$$

hence exact sequences

$$0 \to C^q(\mathscr{U}, F') \to C^q(\mathscr{U}, F) \to C^q(\mathscr{U}, F'') \to 0$$

and hence in the usual way a long exact cohomology sequence

$$0 \to H^0(\mathscr{U}, F') \to H^0(\mathscr{U}, F) \to H^0(\mathscr{U}, F'') \to H^1(\mathscr{U}, F') \to \cdots.$$

The homomorphisms in this sequence commute with the $\theta_{uv}$ hence by taking all proper coverings $\mathscr{U}$ of $X$ we have a direct system of exact sequences. Since the direct limit preserves exactness we have an exact cohomology sequence:

$$(*)\ 0 \to H^0(X, F') \to H^0(X, F) \to H^0(X, F'') \to H^1(X, F') \to \cdots .$$

### 5.1.4

Suppose that $S$ is a Godement sheaf and not only a presheaf. We have defined cohomology groups $H^q(X, S)$ and we have to verify whether axioms $(C_1)$, $(C_2)$ and $(C_3)$ are satisfied:

$(C_1)$ A sheaf homomorphism $S \to T$ induces homomorphisms

$$C^q(\mathscr{U}, S) \to C^q(\mathscr{U}, T)$$

which commute with the coboundary map $\delta$, and induce cohomology homomorphisms

$$H^q(\mathscr{U}, S) \to H^q(\mathscr{U}, T).$$

These cohomology homomorphisms commute with the $\theta_{uv}$ and therefore give rise to limit homomorphisms

$$H^q(X, S) \to H^q(X, T).$$

$(C_2)$ Let $\mathscr{U} = \{U_i\}$ be an open covering of $X$, $f \in C^0(\mathscr{U}, S)$ a 0-cochain, i.e., $f = \{f_i\}$, $f_i \in \Gamma(U_i, S)$.
Then $(\delta f)_{ij} = f_j - f_i$ on $U_i \cap U_j$.

The 0-cochain $f$ is a 0-cocycle if and only if $f_i = f_j$ on $U_i \cap U_j$ for all pairs of indices $i, j$, if and only if the $f_i$ defines a unique global section on $S$ by the axioms $(S_1)$ and $(S_2)$. Since the only 0-coboundary is 0, we have an isomorphism

$$H^0(\mathscr{U}, S) \simeq \Gamma(X, S).$$

By taking the limit on $\mathscr{U}$, we have an isomorphism

$$H^q(X, S) \simeq \Gamma(X, S).$$

$(C_3)$ Let $0 \to S' \to S \to S'' \to 0$ be an exact sequence of sheaves on $X$. For each open subset $U$ of $X$ we have an exact sequence

$$0 \to S'(U) \to S(U) \to S''(U)$$

where the map $S(U) \to S''(U)$ need not be an epimorphism. Let $T$ be the quotient presheaf defined by

$$T(U) = S(U)/S'(U), \quad U \text{ open in } X.$$

We have a presheaf monomorphism $T \to S''$, hence an exact sequence of presheaves

$$(1)\ 0 \to S' \to S \to T'' \to 0,$$
$$(2)\ 0 \to T \to S'' \to S''/T \to 0.$$

The axiom $(C_3)$ will be satisfied if the monomorphism $T \to S''$ induces a cohomology isomorphism,

$$H^q(X, T) \to H^q(X, S'').$$

Indeed, from the exact sequence (1), we have the $(C_3)$.
The isomorphisms are induced from the condition $H^q(X, S''/T) = 0$ for all $q \geq 0$ by the exact sequence (2). Thus $(C_3)$ will be satisfied if the following (5.1.5) holds on the topological space $X$.

### 5.1.5

If $F$ is a presheaf on a paracompact space $X$ such that $\alpha(F) = 0$, then

$$H^q(X, F) = 0 \quad \text{for all} \quad q \geq 0.$$

### Remark 5.1.6

If fact (5.1.5) is not true without some topological restrictions on $X$. We recall that a topological space $X$ is **paracompact** if $X$ is Hausdorff and every open covering of $X$ has a locally finite refinement, namely each point of $X$ has a neighbourhood which meets only a finite number of elements of the refinement.

**Examples 5.1.7**

[1] If $X$ is a locally compact space with a countable base of open sets, in particular manifold, then $X$ is paracompact.

[2] If $X$ is a metric space, then $X$ is paracompact.

**Theorem 5.1.8**

If $X$ is paracompact then (5.1.5) is true.

***Proof.*** Suppose that $F$ is a presheaf on $X$ such that the Cartan sheaf $\alpha(F)$ is the zero sheaf.

Let $\mathscr{U} = \{U_i|_{i\in I}\}$ be an open covering of $X$, and a $q$-cochain $f\in C^q(\mathscr{U}, F)$.

It is sufficient to show that $\mathscr{U}$ admits a refinement $\mathscr{V}=\{V_j\}_{j\in J}$ of $\mathscr{U}$ and $\tau: J\to I$ of the index sets with $V_j \subseteq U_{\tau(j)}$ for all $j\in J$, such that $\tau f = 0$.

Since $X$ is paracompact we may assume that $\mathscr{U}$ is locally finite. From the paracompactness of $X$, there is a refinement $\mathscr{W}=\{W_i\}_{i\in I}$ of $\mathscr{U}$ such that the closure $\overline{W_i}\subset U_i$ for each $i\in I$.

For each point $x\in X$, there is an open neighbourhood $V_x$ of $x$ which meets only a finite number of the $U_i$ in $\mathscr{U}$ with the following conditions:

(1) if $x\in U_i$, then $V_x \subset U_i$, and if $x\in W_i$, then $V_x \subset W_i$,

(2) if $x\notin U_i$, then $V_x\cap W_i = \varnothing$ since $\overline{W_i}\subset U_i$ ,

(3) since $\mathscr{U}$ is locally finite we have $x \in U_{i_0} \cap \cdots \cap U_{i_q}$ for only a finite number of sets $(i_0, \cdots, i_q)$, since $\alpha(F) = 0$, the restriction of $f_{i_0 \cdots i_q}$ to $V_x$ is zero.

Let $\mathscr{V}$ be the covering $\{V_x|_{x \in X}\}$ and choose a map $\tau : X \to I$ such that $x \in W_{\tau(x)} \in \mathscr{W}$.

By (1) $V_x \subset W_{\tau(x)} \subset U_{\tau(x)}$ and so $\mathscr{V}$ is a refinement of $\mathscr{U}$. For each $f_{\tau(x_0) \cdots \tau(x_q)} \in F(U_{\tau(x_0)} \cap \cdots \cap U_{\tau(x_q)})$ the restriction

$(\tau f)_{x_0 \cdots x_q} = \rho_{V_{x_0} \cap \cdots \cap V_{x_q}}^{U_{\tau(x_0)} \cap \cdots \cap U_{\tau(x_q)}} (f_{\tau(x_0) \cdots \tau(x_q)}) \in (V_{x_0} \cap \cdots \cap V_{x_q}).$

If $V_{x_0} \cap \cdots \cap V_{x_q} \neq \varnothing$, then $V_{x_0}$ meets each $V_{x_i} (i = 0, \cdots, q)$, hence meets $W_{\tau(x_i)}$, $i = 0, \cdots, q$.

By (2) $x_0 \in U_{\tau(x_i)}$, $i = 0, \cdots, q$, i.e., $x_0 \in U_{\tau(x_0)} \cap \cdots \cap U_{\tau(x_q)}$.

By (3) the restriction of $f_{\tau(x_0) \cdots \tau(x_q)}$ to $V_{x_0}$ is zero. Hence $(\tau f)_{x_0 \cdots x_q} = 0$ and therefore $\tau f = 0$. □

***Remark 5.1.9***

(1) If $X$ is paracompact then the Čech cohomology groups $H^q(X, S)$ defined above satisfy axioms $(C_1)$, $(C_2)$ and $(C_3)$.

(2) If $S = X \times G$, $G$ a discrete abelian group, $S$ is a constant sheaf, then $H^q(X, S) = H^q(X, G)$ the usual cohomology.

## 5.2 Grothendieck Cohomology

We want to have a sheaf cohomology theory satisfying the axioms $(C_1)$, $(C_2)$ and $(C_3)$ without any restrictions on the topological space $X$: for example for applications to algebraic geometry, in which algebraic varieties are given a topology, namely the Zariski topology, which is not even Hausdorff.

To define the Grothendieck cohomology we recall some homological algebra.

In the category of abelian groups, we characterize the free abelian groups as follows:

## Definition 5.2.1

Suppose that $F$ is an abelian group, $p: G \to G''$ an epimorphism, and $u: F \to G''$ a homomorphism

We have a diagram

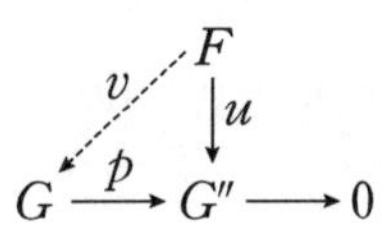

then $F$ is **free** $F$ satisfies the **condition**(P):
given any exact sequence

$$G \xrightarrow{p} G'' \xrightarrow{\alpha} 0$$

and any homomorphism

$$u: F \to G'',$$

there is a homomorphism

$$v: F \to G \quad \text{such that } p \circ v = u.$$

The condition(P) makes sense in any abelian category.
Let $R$ be a ring. The free $R$-modules satisfy the condition(P), but the converse is false.
An object $P$ satisfying condition(P) in an abelian category is called a **projective** object of the abelian category.

Dually, in an abelian category an object $I$ is **injective** if $I$ satisfies the **condition**(I): Given any diagram

$$\begin{array}{ccccc} 0 & \longrightarrow & G' & \xrightarrow{p} & G \\ & & & \searrow^{u} & \downarrow{\scriptstyle v} \\ & & & & I \end{array}$$

in which $p$ is a monomorphism and $u : G' \to I$ a morphism, there is a morphism $v : G \to I$ such that $v \circ p = u$.

***Remark 5.2.2***

In the category of abelian groups, an **injective abelian group** is the same thing as a **divisible abelian group**, i.e., given any element $x$ of the abelian group and any integer $n \neq 0$, there is an element $y$ in the group such that $ny = x$.

**Theorem 5.2.3**

In the category of $R$-modules where $R$ is a ring, the following statements are true.

(P′) Given any object $M$, there is a projective object $P$ and an epimorphism $P \to M$.

(I′) Given any object $M$, there is an injective object $I$ and a monomorphism $M \to I$.

For the proofs of Theorem 5.2.3, one may consult any book on homological algebra.

Let $\mathscr{C}$ be an abelian category, in which (I′) is true. If $M$ is an object of $\mathscr{C}$, then by (I′), we have an exact sequence

$$0 \to M \to I_0 \to M_1 \to 0,$$

where $I_0$ is injective and $M_1 = I_0 / M$.

For the object $M_1$ of $\mathscr{C}$, by (I′), we have an exact sequence

$$0 \to M_1 \to I_1 \to M_2 \to 0,$$

where $I_1$ is injective and $M_2 = I_1 / M_1$, and so on.

By joining these short exact sequences together, we obtain an exact sequence

$$(*) \quad 0 \longrightarrow M \longrightarrow I_0 \xrightarrow{\delta_1} I_1 \xrightarrow{\delta_2} I_2 \longrightarrow \cdots$$

in which the $I$'s are injective.
Such a sequence is called an **injective resolution** of $M$.

**Theorem 5.2.4**

Let $f : M \to M'$ be a morphism in $\mathscr{C}$.

If $I^*(I'^*)$ is an injective resolution of $M$(resp $M'$), then $f$ can be lifted to a chain morphism

$$\{f_n : I_n \to I_n'\}$$

such that

$$f = f_0^* : H^0(I^*) \to H^0(I'^*)$$

and any two such liftings are homotopic.

***Proof.*** We construct $f_n$ by induction on $n$. If this has been done up to $f_{n-1}$ $(n > 0)$ we have a commutative diagram

$$\begin{array}{ccccc} I^{n-1} & \longrightarrow & I^{n-1}/\operatorname{Im}\delta'_{n-2} & \xrightarrow{i} & I^n \\ \downarrow{\scriptstyle f_{n-1}} & & \downarrow{\scriptstyle f'} & & \downarrow{\scriptstyle f_n} \\ I'^{n-1} & \longrightarrow & I'^{n-1}/\operatorname{Im}\delta'_{n-2} & \xrightarrow{j} & I'^n \end{array}$$

in which $i$ and $j$ are injective by the exactness of $I^*$ and $I'^*$, and $f'$ exists since

$$\delta'_{n-1} f_{n-1} \delta_{n-2} = \delta'_{n-1} \delta'_{n-2} f_{n-2} = 0.$$

Hence since $I'^n$ is an injective object, $j \circ f'$ can be extended to a morphism $f_n : I^n \to I'^n$.

The outside square commutes, so $\{f_n\}$ is a morphism of chain complexes.

By the definition of an injective resolution, $M \simeq H^0(I^*)$,

$$M' \simeq H^0(I'^*) \quad \text{and} \quad f = f_0^*.$$

If $\{g_n I^n \to I'^n \mid n \in N\}$ is another obtained map, we can construct a homotopy $\{h_n : I^n \to I'^{n-1} \mid n \in N\}$ by induction.

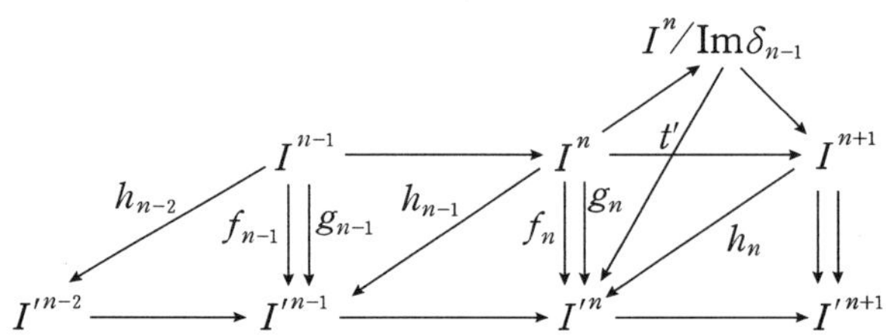


If we assume that $\delta' h_{n-2} + h_{n-1}\delta = f_{n-1} - g_{n-1}$, then the map

$$t = f_n - g_n - \delta'_{n-1} h_{n-1}$$

is such that

$$t\delta = f_n\delta - g_n\delta - \delta' h\delta = f_n\delta - g_n\delta - \delta'(f_n - g_n - \delta h) = 0$$

hence $t$ factors through $I^n/\mathrm{Im}\delta_{n-1}$ as $t'$.

Since $I'^n$ is injective, $t'$ extends to $h_n$ such that

$$h_n\delta + \delta' h_{n-1} = f_n - g_n$$

as required. □

Let $T$ be a covariant additive left exact functor from the abelian category $\mathscr{C}$ to another abelian category $\mathscr{C}'$.
If we operate with $T$ on $(*)$, we have a sequence

$$T(*) \quad 0 \longrightarrow T(M) \longrightarrow T(I_0) \xrightarrow{T(\delta_1)} T(I_1) \xrightarrow{T(\delta_2)} T(I_2) \longrightarrow \cdots,$$

which in general is no longer exact, but since

$$T(\delta_{q+1}) \circ T(\delta_q) = T(\delta_{q+1} \circ \delta_q) = T(0) = 0,$$

$T(*)$ is a complex and hence we have cohomologies,

$$H^q(T(*)) = \mathrm{Ker}\, T(\delta_{q+1}) / \mathrm{Im}\, T(\delta_q), \quad q \geq 0, \quad T(\delta_0) = 0.$$

**Theorem 5.2.5**

The objects $H^q(T(*)) \in \mathscr{C}'$ depend only on $M$ and $T$ and not on the particular injective resolution of $M$ used to construct them.

We denote $R^q T(M)$ for $H^q(T(*))$, and then

$$R^q T : \mathscr{C} \to \mathscr{C}', \quad \text{given by}$$
$$M \mapsto R^q T(M)$$

is a covariant additive functor from $\mathscr{C}$ to $\mathscr{C}'$ for each $q \geq 0$.

Indeed, let $T : \mathscr{C} \to \mathscr{C}'$ be a covariant additive left exact functor between abelian categories, and let $f : M \to N$ be a homomorphism in $\mathscr{C}$.

Left

$$\begin{array}{ccccccccc} 0 \longrightarrow & M & \longrightarrow & I_0 & \xrightarrow{\alpha_1} & I_1 & \xrightarrow{\alpha_2} & I_2 & \longrightarrow \cdots \\ & \downarrow f & & \downarrow f_0 & & \downarrow f_1 & & \downarrow f_2 & \\ 0 \longrightarrow & N & \longrightarrow & I'_0 & \xrightarrow{\alpha'_1} & I'_1 & \xrightarrow{\alpha'_2} & I'_2 & \longrightarrow \cdots \end{array}$$

be injective resolutions and the $f_i$ be induced by $f$.

By applying $T$ for the diagram we have the right derived functor

$$R^d T : \mathscr{C} \to \mathscr{C}', \text{ given by}$$
$$M \mapsto R^d T(M), \text{ for each } q \geq 0.$$

The $R^q T$ is called the **$q$-th right derived functor of $T$.**

Since $T$ is left exact,

$$R^0 T(M) = H^0(T(*))$$
$$= \text{Ker } T(\delta_1) = T(M),$$

and so $R^0 T = T$.

**Theorem 5.2.6**

Let $0 \to M' \to M \to M'' \to 0$ be an exact sequence in $\mathscr{C}$.

Then we have an exact sequence

$$0 \to T(M') \to T(M) \to T(M'') \to R^1 T(M') \to R^1 T(M) \to R^1 T(M'')$$
$$\to R^2 T(M') \to \cdots.$$

The proofs of Theorem 5.2.5 and Theorem 5.2.6 can be found in any book on homological algebra.

We will only consider the category of sheaves of abelian groups on a fixed space $X$, which is an abelian category.

**Theorem 5.2.7**

The (I′) is true for the category of sheaves of abelian groups on $X$.

***Proof.*** We have to show that if $S$ is a sheaf of abelian groups on $X$, then $S$ can be embedded in an injective sheaf.

We assume that Theorem 5.2.3 (I′) is true for the category of abelian groups. Then each $S(x)$ can be embedded in an injective abelian group $J(x)$. Define a presheaf $I$ by the product

$$I(U) = \prod_{x \in U} J(x)$$

for each open $U \subset X$. Then $I$ is a Godement sheaf and for each open $U \subset X$, $S(U)$ is a subgroup of $I(U)$, so that $S$ is a subsheaf of $I$. It remains to show that $I$ is an injective sheaf.

Let $T$ be a sheaf on $X$, $\varphi : T \to I$ a sheaf homomorphism, and $x \in X$. For each neighbourhood $U$ of $x$, we have a homomorphism $I(U) \to J(x)$ given by the projection. By the definition of the direct limit, there is a homomorphism $I(x) \to J(x)$, and hence a homomorphism

$$\psi(x) : T(x) \xrightarrow{\varphi(x)} I(x) \to J(x).$$

Conversely, any family of homomorphisms

$$\psi(x) : T(x) \to J(x), \ x \in X$$

induces a homomorphism $T(U) \to I(U)$.

Hence we have a sheaf homomorphism $T \to I$.

Thus the sheaf homomorphisms $\varphi : T \to I$ are bijective with the families of homomorphisms,

$$\psi(x) : T(x) \to J(x), \ x \in X.$$

Let $T'$ be a subsheaf of $T$ and $\varphi' : T' \to I$ a sheaf homomorphism. By the argument above, $\varphi'$ induces a homomorphism

$$\varphi'(x) : T'(x) \to J(x), \ x \in X.$$

Since $J(x)$ is an injective abelian group, $\psi'(x)$ can be extended to $\psi(x) : T(x) \to J(x)$, and the $\psi(x)$ define a sheaf homomorphism

$\varphi : T \to I$ which extends $\varphi'$.

Hence $I$ is an injective sheaf. □

***Remark 5.2.8***

In general the (P′) is false for the category of sheaves of abelian groups on $X$.

## Definition 5.2.9

The functor $S \mapsto \Gamma(X, S)$ from the category of sheaves on $X$ to the category of abelian groups is covariant, additive, and left exact. By Theorem 5.2.5, it has right derived functors $R^q\Gamma(X, S)$. We define the $q$-th **Grothendieck cohomology group of $X$ with coefficients in $S$** to be

$$H^q(X, S) = R^q\Gamma(X, S),$$

$(C_1)$ is satisfied since $R^q\Gamma$ is a covariant functor.

$(C_2)$ is satisfied since $H^0 = R^0\Gamma = \Gamma$ and $\Gamma$ is left exact.

$(C_3)$ is satisfied by Theorem 5.2.6.

***Remark 5.2.10***

(1) If $X$ is paracompact, then the Grothendieck cohomology groups are the same as the Čech cohomology groups up to isomorphism.

(2) If $X$ is not paracompact, then $H^0$ and $H^1$ in two cohomologies are the same by both definitions, but for $q \geq 2$, the Čech cohomology group $\check{H}^q$ is in general different from the Grothendieck cohomology group $H^q$.

(3) It turns out that there are natural homomorphisms

$$\eta^q : \check{H}^q(X, S) \to H^q(X, S),$$

where $\check{H}^q$ is the Čech group and $H^q$ the Grothendieck group. $\eta^q$ is always an isomorphism for $q=0, 1$ and a monomorphism for $q=2$, and an isomorphism for all $q \geq 0$ if $X$ is paracompact.

**Theorem 5.2.11**

Let $X$ be a topological space, $Y$ a closed subset of $X$, $S$ a sheaf on $Y$, $S^X$ the extension of $S$ to $X$. Then $H^q(X, S^X) = H^q(Y, S)$ for all $q \geq 0$, where $S^X$ is the push down of $S$ on $X$.

***Proof.*** If $I_0$ is an injective sheaf on $Y$, then the extension $I_0^X$ of $I_0$ is an injective sheaf on $X$. Hence if

$$0 \to S \to I_0 \to I_1 \to I_2 \to \cdots$$

is a resolution of $S$ by injective sheaves on $Y$, then

$$0 \to S^X \to I_0^X \to I_1^X \to I_2^X \to \cdots$$

is an injective resolution of $S^X$.

Since $\Gamma(Y, S) = \Gamma(X, S^X)$, it follows that the $H^q(X, S^X)$ and $H^q(Y, S)$ are the cohomology groups of the same complex, hence are equal. □

**Lemma 5.2.12**

If $X$ is an $n$-dimensional paracompact space, and $\mathscr{U}$ an open covering of $X$, then there is a refinement $\mathscr{V}$ of $\mathscr{U}$ such that any $n+2$ distinct open sets in $\mathscr{V}$ have empty intersection.

**Theorem 5.2.13**

Let $X$ be a paracompact space of dimension $n < \infty$, $S$ a sheaf on $X$. Then $H^q(X, S) = 0$ for all integers $q > n$.

***Proof.*** Since $X$ is paracompact we may use the Čech cohomology groups. Let $\mathscr{U}$ be an open covering of $X$. Since $\dim X = n$, by Lemma 5.2.12 there is a refinement $\mathscr{V}$ of $\mathscr{U}$ such that any $n+2$ distinct open sets of $\mathscr{V}$ have empty intersection. It follows that

$$C^q(\mathscr{V};\ S) = 0 \text{ for all } q > n,$$

hence $H^q(\mathscr{V};\ S) = 0$ for all $q > n$. Since such coverings $\mathscr{V}$ are cofinal among all open coverings of $X$, we have $H^q(X, S) = 0$ for all $q > n$. □

TER

# 6 Sheaf on Riemann Surface

## 6.1 Riemann-Roch Theorem

Let $\Sigma$ be a compact connected Riemann surface, $\mathscr{O}$ the sheaf of germs of holomorphic functions on $\Sigma$, $D = \sum_{i=1}^{k} n_i P_i$ a positive divisor (i.e., $n_i > 0$) on $\Sigma$, $\mathscr{O}(D)$ the sheaf of germs of meromorphic functions $f$ such that $(f) + D \geq 0$. Then $\mathscr{O}$ is a subsheaf of $\mathscr{O}(D)$, hence we have a quotient sheaf $S = \mathscr{O}(D)/\mathscr{O}$. If $x \in \Sigma - \{P_1, \cdots, P_k\}$, then $\mathscr{O}(x) = \mathscr{O}(D)(x)$, hence $S(x) = 0$. If $x = P_i$ and $z_i$ is the local coordinate on $\Sigma$ in the neighbourhood of $P_i$ then any element of $\mathscr{O}(D)(P_i)$ is of the form $z_i^{-n_i} g(z_i)$ where $g(z_i)$ is holomorphic, and therefore $S(P_i) \cong \mathbb{C}^{n_i}$, $i = 1, \cdots, k$. Thus $S$ is concentrated on $P_1, \cdots, P_k$. Therefore by Theorem 5.2.11 (here $X = \Sigma$, $Y = \{P_1, \cdots, P_k\}$) and Theorem 5.2.13, we have $H^q(\Sigma, S) = 0$ for all $q > 0$. Hence from the exact sequence

$$0 \to \mathscr{O} \to \mathscr{O}(D) \to S \to 0$$

we have an exact cohomology sequence

$$\begin{aligned} (*) \quad & 0 \to H^0(\Sigma, \mathscr{O}) \to H^0(\Sigma, \mathscr{O}(D)) \to H^0(\Sigma, S) \\ & \to H^1(\Sigma, \mathscr{O}) \to H^1(\Sigma, \mathscr{O}(D)) \to 0. \end{aligned}$$

The elements $H^0(\Sigma, \mathscr{O})$ are the global holomorphic functions on $\Sigma$. By an extension of Liouville's theorem these are constants, since $\Sigma$ is compact and connected, hence $H^0(\Sigma, \mathscr{O}) = \mathbb{C}$

The elements of $H^0(\Sigma, \mathscr{O}(D))$ are the global meromorphic functions $f$ such that $(f) + D \geq 0$.

Since $S$ is concentrated on $P_1, \cdots, P_k$ and $S(P_i) \cong \mathbb{C}^{n_i}$, we have

$$H^0(\Sigma, S) = \Gamma(\Sigma, S) \cong \mathbb{C}^{\deg(D)}.$$

In fact the first cohomology $H^1(\Sigma, \mathscr{O})$ is a finite dimensional vector space over $\mathbb{C}$. We will see this finiteness at Theorem 7.2.6 of Chapter 7. We denote

$$\dim_{\mathbb{C}} H^1(\Sigma, \mathcal{O}) = g$$

which is called the **genus** of $\Sigma$. Also $H^1(\Sigma, \mathcal{O}(D))$is a finite dimensional vector space over $\mathbb{C}$, since from the exact sequence it is a homomorphic image of $H^1(\Sigma, \mathcal{O})$ and all the homomorphisms in the ( $*$) are $\mathbb{C}$-linear. We denote

$$\dim_{\mathbb{C}} H^1(\Sigma, \mathcal{O}(D)) := i(D)$$

called the **index of speciality** of $D$. By counting up the dimensions in the exact sequence ( $*$), we have

$$1 - \dim_{\mathbb{C}} H^0(\Sigma, \mathcal{O}(D)) + \deg(D) - g + i(D) = 0.$$

Thus the number of linearly independent meromorphic functions $f$ on $\Sigma$ such that $(f) + D \geq 0$ is

$$\dim_{\mathbb{C}} H^0(\Sigma, \mathcal{O}(D)) = \deg(D) + 1 - g + i(D) \text{ , and}$$
$$0 \leq i(D) \leq g.$$

This is one of the **Riemann-Roch Theorem**.

In particular, if $\deg(D) \geq g$, then there are global meromorphic functions $f$ such that $(f) + D \geq 0$, i.e., such that the poles of $f$ are contained in $D$.

## 6.2 Divisor Class Group

Let $\Sigma$ be a compact connected Riemann surface, $\mathcal{O}^*$ the sheaf of germs of non-vanishing holomorphic functions and the group operation in the stalks of $\mathscr{O}^*$ is multiplication, $\mathscr{M}^*$ the sheaf of germs of meromorphic functions on $\Sigma$ which do not vanish identically and the group operation in the stalks of $\mathscr{M}^*$ is multiplication. Then $\mathscr{O}^*$ is a subsheaf of $\mathscr{M}^*$. Then we have an exact sequence

$$0 \to \mathscr{O}^* \to \mathscr{M}^* \to \mathscr{D} \to 0$$

where $\mathscr{D} = \mathscr{M}^* / \mathscr{O}^*$ is the **sheaf of germs of divisors** on $\Sigma$. If $U$ is an open subset $\Sigma$, then the elements of $\mathscr{D}(U)$ are in 1-1 correspondence with the divisors on $U$.

In particular $H^0(\Sigma, \mathscr{D}) = \mathscr{D}(\Sigma)$ is the group of all divisors on $\Sigma$. We have an exact cohomology sequence

$$0 \to H^0(\Sigma, \mathscr{O}^*) \to H^0(\Sigma, \mathscr{M}^*) \to H^0(\Sigma, \mathscr{D}) \xrightarrow{\delta} H^1(\Sigma, \mathscr{O}^*) \to H^1(\Sigma, \mathscr{M}^*) \to \cdots$$

and $\delta$ can be shown to be an epimorphism, so that

$$H^1(\Sigma, \mathscr{O}^*) \cong H^0(\Sigma, \mathscr{D}) / \mathrm{Im}\, H^0(\Sigma, \mathscr{M}^*).$$

$H^0(\Sigma, \mathscr{D})$ is the group of all divisors on $\Sigma$, and $\mathrm{Im}\, H^0(\Sigma, \mathscr{M}^*)$ is the group of all principal divisors $(f)$ on $\Sigma$. Hence $H^1(\Sigma, \mathscr{O}^*)$ is the group of all classes of divisors modulo principal divisors, we call it the **divisor class group** $\Delta(\Sigma)$ of $\Sigma$.

## 6.3 Jacobian

Let $\Sigma$, $\mathscr{O}$, $\mathscr{O}^*$ have the same meaning as in section 6.2. If $f$ is holomorphic then $\exp(2\pi\, if)$ is non-vanishing holomorphic, hence we have a homomorphism of sheaves

$$\exp(2\pi i) : \mathscr{O} \to \mathscr{O}^*.$$

This is an epimorphism, and since $\Sigma$ is connected its kernel is the integers $\mathbb{Z}$, the constant sheaf of integers.

Thus we have an exact sequence of sheaves

$$0 \to \mathbb{Z} \to \mathscr{O} \xrightarrow{\exp 2\pi i} \mathscr{O}^* \to 0$$

and an exact cohomology sequence

$$0 \to H^0(\Sigma, \mathbb{Z}) \to H^0(\Sigma, \mathscr{O}) \xrightarrow{\exp 2\pi i} H^0(\Sigma, \mathscr{O}^*) \xrightarrow{\delta} H^1(\Sigma, \mathbb{Z}) \to \cdots.$$

Since $\Sigma$ is connected, $H^0(\Sigma, \mathbb{Z}) = \mathbb{Z}$,

$$H^0(\Sigma, \mathscr{O}) = \mathbb{C}(\text{Liouville theorem}),$$

$H^0(\Sigma, \mathscr{O}^*) = \mathbb{C} - \{0\}$(the multiplicative group of nonzero complex numbers). Hence $exp 2\pi i$ is an epimorphism in the cohomology exact sequence, and $\delta = 0$. We have an exact sequence,

$$(*)\ 0 \to H^1(\Sigma, \mathbb{Z}) \to H^1(\Sigma, \mathscr{O}) \to H^1(\Sigma, \mathscr{O}^*) \xrightarrow{\delta} H^2(\Sigma, \mathbb{Z}) \to \cdots,$$

where $H^1(\Sigma, \mathscr{O}^*) = \Delta(\Sigma)$ is the divisor class group of $\Sigma$, and $H^2(\Sigma, \mathbb{Z}) = \mathbb{Z}$ since $\Sigma$ is compact and connected.

Consider the homomorphism $\delta$. Let $D$ be a divisor on $\Sigma$, and triangulate $\Sigma$ so that the points $P$ which occur in $D$ with non-zero coefficients lie in the interior of the 2-simplexes of the triangulation and so that at most one of the points $P$ lies in any 2-simplex.

Let $\mathscr{U}$ be the open covering of $\Sigma$ formed by the stars $U_i$ of the vertices $A_i$ of the triangulation. Let $P$ be one of the points which occurs in $D$ with a non-zero coefficient, and suppose $P$ lies in the simplex $A_iA_jA_k$. Choose meromorphic functions $h_i$ on $U_i$ such that $(h_i) = D|_{U_i}$, then $f_{ij} = h_i h_j^{-1}$ is a 1-cocycle of $\mathscr{U}$ with values in $\mathscr{O}^*$, whose image in $H^1(\Sigma, \mathscr{O}^*) = \Delta(\Sigma)$ is the class $\{D\}$ of $D$. Hence $\delta\{D\}$ is represented by the 2-cocycle $v \in C^2(\mathscr{U}, \mathbb{Z})$, where

$$
\begin{aligned}
v_{ijk} &= \frac{1}{2\pi i}(\log f_{jk} - \log f_{ik} + \log f_{ij}) \text{ on } U_i \cap U_j \cap U_k \\
&= \text{total variation of } \log h \text{ on going round the triangle } A_i A_j A_k \\
&= \text{zeros-poles in } U_i \cap U_j \cap U_k \\
&= \deg(D|_{U_i \cap U_j \cap U_k}),
\end{aligned}
$$

hence $\delta(D) = \deg D$.

Thus the $\delta$ maps a divisor class to its degree, and the kernel of $\delta$ is the subgroup $J(\Sigma)$ of $\Delta(\Sigma)$ consisiting of all divisor classes of degree 0. The $J(\Sigma)$ is called the **Jacobian** of $\Sigma$. Considering the exact sequence $(*)$, we have an exact sequence

$$
0 \to H^1(\Sigma, \mathbb{Z}) \to H^1(\Sigma, \mathscr{O}) \to J(\Sigma) \to 0,
$$
$$
\text{i.e., } J(\Sigma) \cong H^1(\Sigma, \mathscr{O}) / H^1(\Sigma, \mathbb{Z}).
$$

Since $H^1(\Sigma, \mathscr{O}) \cong \mathbb{C}^g \cong \mathbb{R}^{2g}$ and $H^1(\Sigma, \mathbb{Z}) \cong \mathbb{Z}^{2g}$, the Jacobian of $\Sigma$ is a complex torus of complex dimension $g$,

$$
\text{i.e., } J(\Sigma) = \mathbb{R}^{2g} / \mathbb{Z}^{2g} = T^g.
$$

***Remark 6.3.1***

(1) The divisor class group

$$
\mathrm{Pic}(\Sigma) := H^1(\Sigma, \mathscr{O}^*)
$$

is called the **Picard group** of $\Sigma$, which is a 1-1 correspondence with the set of holomorphic line bundles on $\Sigma$.

(2) The second cohomology group

$$
H^2(\Sigma, \mathbb{Z}) \cong \mathbb{Z}
$$

is a 1-1 correspondence with the set of complex line bundles on $\Sigma$

via their first Chern classes.

(3) In Riemann surface each complex bundle on $\Sigma$ yields the set of holomorphic line bundles parametrized by the Jacobian $J(\Sigma)$ of $\Sigma$.

## 6.4 Differential

Let $\Sigma$ be a Riemann surface, $\mathscr{U}=\{U_i\}$ an open covering of $\Sigma$ of coordinate neighbourhood, $z_i$ the local coordinate in $U_i$ and let $\varphi_i$ be a holomorphic function on $U_i$ such that

$$\varphi_i dz_i = \varphi_j dz_j,$$

i.e., $dz_j/dz_i = \varphi_i \varphi_j^{-1}$ on $U_i \cap U_j$.

Then $w=\{\varphi_i\}$ is called a **holomorphic differential** on $\Sigma$, or a **holomorphic covariant vector field** on $\Sigma$.

Since $\dfrac{dz_j}{dz_i}$ is holomorphic and non-zero in $U_i \cap U_j$, we have

$$(\varphi_i)|_{U_i \cap U_j} = (\varphi_j)|_{U_i \cap U_j}$$

and hence the local divisors on $U_i$ can be fitted together to form a global divisor $(w)$, the divisor of the holomorphic differential $w$. Since $w$ is holomorphic, $(w) \geq 0$.

If $f$ is a holomorphic function on $\Sigma$, then

$$df = \left\{ \frac{df}{dz_i} \right\}$$

is a holomorphic differential on $\Sigma$. The set of all holomorphic differentials on $\Sigma$ is an additive group $\Omega^1(\Sigma)$ and $f \mapsto df$ is a homomorphism

$$d : \Gamma(\Sigma, \vartheta) \to \Omega^1(\Sigma).$$

If $U$ is an open subset of $\Sigma$ then the above definitions apply to $U$ instead of $\Sigma$, hence we have a group $\Omega^1(U)$ of holomorphic differential on $U$, and as $U$ runs through the open subsets of $\Sigma$ the groups $\Omega^1(U)$ form a sheaf $\Omega^1$, the **sheaf of germs of holomorphic differentials** on $\Sigma$.

If $D$ is a positive divisor on $\Sigma$ then by considering those holomorphic differentials $w$ on $\Sigma$ for which $(w)-D\geq 0$ we have a sheaf $\Omega^1(-D)$ of holomorphic differentials which vanish on $D$.

The differential operator $d$ gives rise to a sheaf holomorphism $d:\mathcal{O}\to\Omega^1$. $d$ is an epimorphism. Indeed, it is enough to verify this locally, that is in effect when $\Sigma=\mathbb{C}$, then it is obvious. The kernel of $d$ consists of the holomorphic function $f$ such that $df=0$. i.e., $\mathrm{Ker}(d)=\mathbb{C}$. Hence we have an exact sequence of sheaves

$$0\to\mathbb{C}\to\mathcal{O}\to\Omega^1\to 0$$

and an exact cohomology sequence

$$0\to H^0(\Sigma,\mathbb{C})\xrightarrow{\alpha} H^0(\Sigma,\mathcal{O})\xrightarrow{\beta} H^0(\Sigma,\Omega^1)\to\cdots.$$

From now on we assume that $\Sigma$ is compact and connected.

Then $H^0(\Sigma,\mathbb{C})=\mathbb{C}$, $H^0(\Sigma,\mathcal{O})=\mathbb{C}$ by Liouville theorem. Hence $\alpha$ is an isomorphism and $\beta=0$.

We have an exact sequence

$$0\to H^0(\Sigma,\Omega^1)\to H^1(\Sigma,\mathbb{C})\to H^1(\Sigma,\mathcal{O})\to H^1(\Sigma,\Omega^1)\xrightarrow{\delta} H^2(\Sigma,\mathbb{C})\to\cdots.$$

In fact $H^1(\Sigma,\Omega^1)\cong\mathbb{C}$, $H^2(\Sigma,\mathbb{C})\cong\mathbb{C}$ and $\delta$ an isomorphism. Hence we have an exact sequence

$$0\to H^0(\Sigma,\Omega^1)\to H^1(\Sigma,\mathbb{C})\to H^1(\Sigma,\mathcal{O})\to 0.$$

If $b_1$ is the 1st Betti number of $\Sigma$, then

$$(*) \quad \dim_{\mathbb{C}} H^0(X, \Omega^1) - b_1 + g = 0,$$

where $g$ is the genus of $\Sigma$.

Let $\mathscr{U} = \{U_i\}$ be an open covering of $\Sigma$, $D$ a positive divisor.

Let $f$ be a 1-cocycle of $\mathscr{U}$ with values in $\vartheta(D)$ and $g$ a 0-cocycle of $\mathscr{U}$ with values in $\Omega^1(-D)$. Define $h$ by

$$h_{ij} = f_{ij} g_i \text{ on } U_i \cap U_j.$$

Since $(f_{ij}) + D \geq 0$ and $(g_i) - D \geq 0$ on $U_i \cap U_j$, we have $(h_{ij}) \geq 0$ on $U_i \cap U_j$, so that $h$ is holomorphic, i.e., $h \in C^1(\mathscr{U}, \Omega^1)$.

Also we have

$$\begin{aligned}(\delta h)_{ijk} &= h_{jk} - h_{ik} + h_{ij} \\ &= f_{jk} - f_{ik} g_i + f_{ij} g_i \\ &= f_{jk} (\delta g)_{ij} - (\delta f)_{ijk} g_i = 0\end{aligned}$$

since $f$ and $g$ are cocycles.

Similarly if $f$ a coboundary, then $h$ is a coboundary.

Hence the pairing $(f, g) \mapsto h$ induces a pairing

$$H^1(\mathscr{U}, \mathscr{O}(D)) \bigotimes_{\mathbb{C}} H^0(\mathscr{U}, \Omega^1(-D)) \to H^1(\mathscr{U}, \Omega^1)$$

and by taking the direct limit a pairing

$$H^1(\Sigma, \mathscr{O}(D)) \bigotimes_{\mathbb{C}} H^0(\Sigma, \Omega^1(-D)) \to H^1(\Sigma, \Omega^1) \equiv \mathbb{C}.$$

It is a special case of a general theorem due to Sarre that this pairing is a duality, i.e., it induces an isomorphism of $H^1(\Sigma, \mathscr{O}(D))$ onto the dual of $H^0(\Sigma, \Omega^1(-D))$. In particular, taking $D = 0$, $H^0(\Sigma, \Omega)$ and $H^1(\Sigma, \mathscr{O})$ have the same dimension so that

$$\dim_{\mathbb{C}} H^0(\Sigma, \Omega^1) = g.$$

**Theorem 6.4.1**

(1) The number of linearly independent holomorphic differentials on $\Sigma$ is $g$.

(2) From equation $(*)$ we have the first Betti number $b_1 = 2g$.

(3) In the Riemann-Roch theorem, the index of speciality

$$i(D) = \dim_{\mathbb{C}} H^1(\Sigma, \mathscr{O}(D)).$$

By Serre's duality theorem we have

$i(D) = \dim_{\mathbb{C}} H^0(\Sigma, \Omega^1(-D))$

$=$ the number of linearly independent holomorphic differentials which vanish on $D$.

***Remark 6.4.2***

In Theorem 6.4.1(2), our definition of the genus agrees with the usual topological genus.

## 6.5 Laurent Theorem

If $\Sigma = \mathbb{CP}^1$ is the Riemann sphere, then by Theorem 6.4.1(2) the 1st cohomology group $H^1(\mathbb{CP}^1, \mathscr{O}) = 0$.

Let $0 < r < R$ be positive real numbers. Let $U_0 = \{z \in \mathbb{CP}^1 : |z| < R\}$, $U_\infty = \{z \in \mathbb{CP}^1 : |z| > r\}$. Then $\mathscr{U} = \{U_0, U_\infty\}$ is an open covering of $\mathbb{CP}^1$, and $U_0 \cap U_\infty$ is the annulus $\{z \in \mathbb{CP}^1 : r < |z| < R\}$. Let $f = f_{0\infty}$ be a 1-cocycle of $\mathscr{U}$ with coefficients in $\mathscr{O}$, i.e., $f$ is holomorphic in $U_0 \cap U_\infty$. Then

$H^1(\mathscr{U}, \mathscr{O}) = 0 \Leftrightarrow f$ is coboundary

$$\Leftrightarrow \ f = (\delta g)_{0\infty} = g_\infty - g_0,$$

where $g_\infty$ is holomorphic in $U_\infty$, and $g_0$ is holomorphic in $U_0$.

This equivalence is called the **Laurent theorem**.

Since $H^1(\mathbb{CP}^1, \mathscr{O}) = H^1(\mathscr{U}, \mathscr{O})$, Laurent theorem is equivalent to the condition $H^1(\mathbb{CP}^1, \mathscr{O}) = 0$.

## 6.6 Weierstrass Theorem

Let $\Sigma$ be a connected compact Riemann surface of genus $g$, $P_0 \in \Sigma$, $\mathscr{O}$ the sheaf of germs of holomorphic functions on $X$ and $\mathscr{O}_0$ the sheaf of germs of meromorphic functions on $\Sigma$ which are holomorphic except at $P_0$.

Then $\mathscr{O}$ is a subsheaf of $\mathscr{O}_0$ and the quotient sheaf $S = \mathscr{O}_0 / \mathscr{O}$ is concentrated on $P_0$. Thus we have an exact sequence

$$0 \to H^0(\Sigma, \mathscr{O}) \to H^0(\Sigma, \mathscr{O}_0) \to H^0(\Sigma, S) \to H^1(\Sigma, \mathscr{O}) \to H^1(\Sigma, \mathscr{O}_0),$$

$H^0(\Sigma, S) := V$ is the additive group of principal parts at $P_0$

$=$ the vector space over $\mathbb{C}$ generated by $z_0^{-1}, z_0^{-2}, \cdots,$

where $z_0$ is the local coordinate at $P_0$.

$\operatorname{Im} H^0(\Sigma, \mathscr{O}_0) := W$ is the subspace of $V$ consisting of principal parts at $P_0$ of global meromorphic functions which are holomorphic except at $P_0$

Since $H^1(\Sigma, \mathscr{O}_0) = 0$ we have

$$V / W \cong H^1(\Sigma, \mathscr{O}) \cong \mathbb{C}^g.$$

**Theorem 6.6.1** **Weierstrass**

Let $\Sigma$ be a connected compact Riemann surface with genus $g$, and $P_0 \in \Sigma$.

Then there is $g$ linearly independent principal parts at $P_0$ which are not principal parts of a global meromorphic function, in other words, there are exactly $g$ positive integers $n_1, \cdots, n_g$ such that there is no global meromorphic function with a pole of order $n_i$ at $P_0$ and no poles elsewhere.

TER

7

# Complex Manifold

## Definition 1.

A complex valued function $f : \mathbb{C}^n \to \mathbb{C}$ is **holomorphic at a point** of $\mathbb{C}^n$ if $f$ can be expanded as a convergent power series in some neighbourhood of the point, and holomorphic on $U \subset \mathbb{C}^n$ if it is **holomorphic** at every point of $U$. Let $\mathscr{O}_{\mathbb{C}^n}$ denote the sheaf of germs of holomorphic functions on $\mathbb{C}^n$.

## Definition 2.

A **complex manifold of dimension** $n$ is a pair $(X, \mathscr{O}_X)$ where $X$ is a second countable Hausdorff topological space and $\mathscr{O}_X$ is a subsheaf of the sheaf of germs of continuous functions on $X$, such that $(X, \mathscr{O}_X)$ is locally isomorphic to $(\mathbb{C}^n, \mathscr{O}_{\mathbb{C}^n})$.

As an example, a complex manifold of dimension 1 is the same thing as a Riemann surface.

From Definition 2 we can introduce local coordinates $z_1, \cdots, z_n$ in the neighbourhood of any point of a complex manifold of dimension $n$.

## Definition 3.

Let $X$ be a complex manifold of dimension $n$. $Y$ is a **complex submanifold** of $X$ of dimension $m$ if the pair $(X, Y)$ is locally isomorphic to the pair $(\mathbb{C}^n, \mathbb{C}^m)$, that is, if $y \in Y$ then we can choose local coordinates $z_1, \cdots, z_n$ on $X$ in the neighbourhood of $y$ such that the first $m$ coordinates $z_1, \cdots, z_m$ are local coordinates on $Y$.

## 7.1 Compact Complex Manifold

The complex projective space $\mathbb{CP}^n$ and any algebraic submanifolds of $\mathbb{CP}^n$ are compact complex manifolds.

Any closed complex manifold of $\mathbb{CP}^n$ is algebraic.

Any compact Riemann surface is algebraic, i.e., is isomorphic to an algebraic submanifold of $\mathbb{CP}^n$ for some $n$. However, a compact complex manifold of dimension $n > 1$ need not be algebraic.

**Theorem 7.1.1**

(1) **Kodaira Theorem** : A compact complex manifold is projective algebraic if and only if it is a Hodge manifold.

(2) **Chow Theorem** : A projective algebraic manifold is algebraic.

**Example 7.1.2**

Let $X$ be a complex manifold of complex dimension $n$, let $\varepsilon^{p,q}$ be the sheaf of $(p, q)$ forms on $X$, and consider the sequence of sheaves, for $p \geq 0$, fixed,

$$0 \to \Omega^p \xrightarrow{i} \varepsilon^{p,0} \xrightarrow{\bar{\partial}} \varepsilon^{p,1} \xrightarrow{\bar{\partial}} \cdots \to \varepsilon^{p,n} \to 0,$$

where $\Omega^p$ is defined as the kernel sheaf of the mapping $\varepsilon^{p,0} \xrightarrow{\bar{\partial}} \varepsilon^{p,1}$, which is the sheaf of **holomorphic differential forms of type** $(p, 0)$ (and we usually say **holomorphic forms of degree** $p$); i.e., in local coordinates, $\varphi \in \Omega^p(U)$ if and only if

$$\varphi = \sum_{|I|=p} \varphi_I \, dz^I, \quad \varphi_I \in \mathcal{O}(U), \quad dz^I = dz_{i_1} \wedge \cdots \wedge dz_{i_p},$$

and we note that $\Omega^0 = \mathcal{O} (= \mathcal{O}_X)$ . Then for each $p$ we have a differential sheaf

$$0 \rightarrow \Omega^p \rightarrow \varepsilon^{p,*}$$

since $\bar{\partial}^2 = 0$, which is, in fact, a resolution of the sheaf $\Omega^p$, by virtue of the Grothendieck version of the Poincaré lemma (see Theorem 8.3.4) for the $\bar{\partial}$-operator. Namely, if $\omega$ is a $(p, q)$-form defined in a polydisc $\Delta$ in $\mathbb{C}^n$, $\Delta = \{z : |z_i| < r,\ i = 1, \cdots, n\}$, and $\bar{\partial}\omega = 0$ in $\Delta$, then there exists a $(p, q-1)$-form $u$ defined in a slightly smaller polydisc $\Delta' \subset \Delta$, so that $\bar{\partial} u = w$ in $\Delta'$.

**Example 7.1.3**

Let $X$ be a complex manifold and consider the differential sheaf over $X$,

$$0 \rightarrow \mathbb{C} \rightarrow \Omega^0 \xrightarrow{\partial} \Omega^1 \rightarrow \cdots \xrightarrow{\partial} \Omega^n \rightarrow 0,$$

where the $\Omega^p$ are defined in Example 7.1.2. Then we claim that this is a resolution of the constant sheaf $\mathbb{C}$. First we note that $\partial = d$, when acting on holomorphic forms of degree $p$, since $d = \partial + \bar{\partial}$, and $\bar{\partial}(\Omega^p) = 0$ for $p = 0, \cdots, n$; then exactness at $\Omega^0$ is immediate.
Moreover, one can locally solve the equation $\partial u = \omega$ for $u$ if $\partial\omega = 0$ by the same type of the proof as for the operator $\bar{\partial}$ indicated in Example 7.1.2.

Suppose that $\mathscr{L}^*$ and $\mathscr{M}^*$ are differential sheaves. Then a **homomorphism** $f : \mathscr{L}^* \rightarrow \mathscr{M}^*$ is a sequence of homomorphisms $f_j : \mathscr{L}^j \rightarrow \mathscr{M}^j$ which commutes with the differentials of $\mathscr{L}^*$ and $\mathscr{M}^*$. Similarly, a **homomorphism of resolutions** of sheaves

$$\begin{array}{ccccc} 0 & \rightarrow & \mathscr{A} & \rightarrow & \mathscr{A}^* \\ & & \downarrow & & \downarrow \\ 0 & \rightarrow & \mathscr{C} & \rightarrow & \mathscr{C}^* \end{array}$$

is a homomorphism of the underlying differential sheaves.

**Theorem 7.1.4**

Let $\varphi \in \varepsilon^{p,q}(U)$ for $U$ open be 1 in $\mathbb{C}^n$ and suppose that $d\varphi = 0$. Then for any point $p \in U$ there is a neighbourhood $N$ of $p$ and a differential form $\Psi \in \varepsilon^{p-1,q-1}(N)$ such that

$$\partial\bar{\partial}\Psi = \varphi \quad \text{in} \quad N.$$

***Proof.*** The proof consists of an application of the Poincaré lemmas for the operators $d$, $\partial$, and $\bar{\partial}$. Namely, since $d\varphi = 0$, we have that there is a $u \in \varepsilon_x^{r-1}$(using germs at $x$), so that $du = \varphi$, where $r = p+q$ is the total degree of $\varphi$. Thus we see that if we write $u = u^{r-1,0} + \cdots + u^{0,r-1}$, we have

$$du = \bar{\partial}u^{p,q-1} + \partial u^{p-1,q}$$
$$\bar{\partial}u^{p-1,q} = \partial u^{p,q-1} = 0,$$

and then there exist (Poincaré lemma for $\partial$ and $\bar{\partial}$) forms $\Psi_1 \in \varepsilon_x^{p-1,q-1}$ and $\Psi_2 \in \varepsilon_x^{p-1,q-1}$ such that

$$\partial\Psi_1 = u^{p,q-1}$$
$$\bar{\partial}\Psi_2 = u^{p-1,q}$$

which implies that

$$\begin{aligned}\varphi = du &= \bar{\partial}\partial\Psi_1 + \partial\bar{\partial}\Psi_2 \\ &= \partial\bar{\partial}(\Psi_2 - \Psi_1).\end{aligned}$$

□

***Remark 7.1.5***

Let $\mathscr{H} := \operatorname{Ker}(\partial\bar{\partial} : \varepsilon^{0,0} \to \varepsilon^{1,1})$ on a complex manifold $X$. Then there is a resolution

$$0 \longrightarrow \mathscr{H} \longrightarrow \varepsilon^{0,0} \xrightarrow{\partial\bar{\partial}} \varepsilon^{1,1} \xrightarrow{d} \varepsilon^{2,1} \oplus \varepsilon^{1,2} \longrightarrow \cdots$$

where the $\mathscr{H}$ is called the sheaf of **pluriharmonic** functions. Theorem 7.1.4 shows exactness at the $\varepsilon^{1,1}$ term.

## 7.2 Stein Manifold

**Definition 7.2.1**

Let $X$ be a complex manifold. Let $\mathscr{H}(X) = \{f : X \to \mathbb{C} : \text{holomorphic}\}$ be the set of all global holomorphic functions on $X$, a subset $Y \subset X$. The **holomorphic envelope** of $Y$ in $X$ is the set

$$H(Y) = \{x \in X \mid f \in \mathscr{H}(X) \text{ implies } |f(x)| \leq \sup_{y \in Y} |f(y)|\}.$$

If $H(Y) = Y$ then $Y$ is said **holomorphically convex** in $X$.
For example, a domain in $\mathbb{C}^1$ is holomorphically convex.

***Remark 7.2.2***

The definition of the **convex hull** of a subset $Y$ of $\mathbb{R}^n$ is the same one if we replace holomorphic functions with linear functions.

**Definition 7.2.3**

A complex manifold $X$ of dimension $n$ is a **Stein manifold** if the following three conditions hold:

(1) $\mathscr{H}(X)$ separates the points of $X$, i.e., if $x, y \in X$, $x \neq y$, then $f(x) \neq f(y)$ for some $f \in \mathscr{H}(X)$.

(2) $\mathscr{H}(X)$ provides local coordinates at each point of $X$, i.e., if $x \in X$, then there are $f_1, \cdots, f_n \in \mathscr{H}(X)$ such that the $f_i$ are local coordinates at $x$.

(3) If $K$ is a compact subset of $X$, then the holomorphic envelope $\mathscr{H}(K)$ of $K$ is compact.

Any domain in $\mathbb{C}^n$ satisfies (1) and (2), but not necessarily (3). A domain in $\mathbb{C}^n$ satisfying (3) is a **domain of holomorphy**.

A closed submanifold of a Stein manifold is a Stein manifold, in particular a closed submanifold of $\mathbb{C}^n$ is a Stein manifold.

Conversely every Stein manifold is isomorphic to a closed submanifold of $\mathbb{C}^n$ for some $n$.

Let $X$ be a complex manifold. The sheaf $\mathscr{O}_X$ is a sheaf of rings if and only if for each open $U \subset X$, $\mathscr{O}_X(U)$ is a ring and for $V \subset U$ the restrictions $\mathscr{O}_X(U) \rightarrow \mathscr{O}_X(V)$ are ring homomorphisms.

### Definition 7.2.4

An **analytic sheaf** on $X$ is a sheaf $S$ of abelian groups such that each $S(U)$ is an $\mathscr{O}_X(U)$-module and for $V \subset U$ the restrictions $S(U) \rightarrow S(V)$ are compatible with the ring homomorphism $\mathscr{O}_X(U) \rightarrow \mathscr{O}_X(V)$.
If we regard $S$ as a Cartan sheaf, then each stalk $S(x)$ is an $\mathscr{O}_X(x)$-module and the operation of $\mathscr{O}_X$ on $S$ is continuous.
For example, $\mathscr{O}_X^p$(=the direct sum of $p$ copies of $\mathscr{O}_X$) is an analytic sheaf. It is defined by

$$\mathscr{O}_X^p(U) = \bigoplus^p \mathscr{O}_X(U),$$

which is the direct sum of $p$ copies of $\mathscr{O}_X(U)$.

## Definition 7.2.5

Let $S$ be an analytic sheaf on a complex manifold $X$. The sheaf $S$ is said a **coherent** if $S$ is locally isomorphic to the cokernel of a homomorphism $\mathcal{O}_X^p \to \mathcal{O}_X^q$, i.e., if $S$ is locally finitely generated and finitely related.

In other words, each $x \in X$ has an open neighbourhood $U$ and an exact sequence

$$\mathcal{O}_X^p|_U \to \mathcal{O}_X^q|_U \to S|_U \to 0$$

where $p$ and $q$ depend on the point $x \in X$.

Thus $\mathcal{O}_X^p(p=1, 2, \cdots)$ is a coherent analytic sheaf.

**Theorem 7.2.6**

Let $X$ be a Stein manifold, $S$ a coherent analytic sheaf on $X$. Then

(1) Each stalk of $S(x)$ is generated, as an $\mathcal{O}_X(x)$-module, by $H^0(X, S)$.

(2) $H^q(X, S)=0$ for all $q>0$.

**Theorem 7.2.7**

Let $X$ be a compact complex manifold, $S$ a coherent analytic sheaf on $X$. Then each $H^q(X, S)$ is a finite dimensional vector space over $\mathbb{C}$.

**Theorem 7.2.8** **Leray**

Let $X$ be a topological space and $\mathcal{U}=\{U_i\}$ an open covering of $X$. Suppose that for all sets of indices $i_0, \cdots, i_q$, we have $H^q(U_{i_0} \cap \cdots \cap U_{i_q}, S)=0$ for all $q>0$.

Then $H^q(X, \mathcal{S}) \cong H^q(\mathscr{U}, \mathcal{S})$ for all $q > 0$.

Theorem 7.2.8 is a consequence of Leray's spectral sequence for an open covering $\mathscr{U}$, which relates the Čech groups $H^q(\mathscr{U}, \mathcal{S})$ and the Grothendieck cohomology groups $H^q(X, \mathcal{S})$.

**Theorem 7.2.9** **Montel**

Let $D$ be a domain in $\mathbb{C}^m$, $\{f_n\}$ a sequence of holomorphic functions in $D$ such that $|f_n| < K$ for all $n$ at all points of $D$.
Then the sequence $\{f_n\}$ has a subsequence which is uniformly convergent on all compact subsets of $D$, and therefore converges to a holomorphic function on $D$.

We translate Theorem 7.2.9 into the language of topological vector spaces. A linear map $u : E \to F$ between topological vector spaces $E$ and $F$ is **compact** if there is a neighbourhood $V$ of the origin $O$ in $E$ such that $\overline{u(V)}$ is compact.

If $D$ is a domain in $\mathbb{C}^m$, then $\mathscr{H}(D)$ is a complex vector space and we give $\mathscr{H}(D)$ the compact-open topology. Let $K$ be a compact subset of $D$, $\delta$ a positive real number, and $V(K, \delta)$ the set of all $f \in \mathscr{H}(D)$ such that $|f(x)| < \delta$ for all $x \in K$. Then the sets $V(K, \delta)$ form a base for the neighbourhoods of $O$ in $\mathscr{H}(D)$. Then $\mathscr{H}(D)$ is a locally convex Hausdorff topological vector space.

Moreover it is metrizable and complete, i.e., $\mathscr{H}(D)$ is a Fréchet space.

**Theorem 7.2.10**

Let $D$ and $D'$ be domains in $\mathbb{C}^n$ such that the closure $\overline{D}$ is a compact subset of $D'$. Then the restriction map $\rho : \mathscr{H}(D') \to \mathscr{H}(D)$ is compact.

***Proof.*** Let $V = \{f \in \mathscr{H}(D') : |f(z)| < 1$ for all $z \in \overline{D}\}$, then $V$ is a neighbourhood of $O$ in $\mathscr{H}(D')$. Then $\rho(V)$ is formed of functions which are uniformly bounded in $D$.

By Theorem 7.2.9, every sequence of elements of $\rho(V)$ has a convergent subsequence. Hence the closure $\overline{\rho(V)}$ is compact and therefore $\rho$ is compact. □

**Theorem 7.2.11** **Schwarz**

Let $E$ and $F$ be Fréchet spaces, and

$u, v: E \to F$ continuous linear maps. Suppose $u$ is an epimorphism and $v$ is compact. Then $\mathrm{Coker}(u+v)$ is finite dimensional.

***Proof of Theorem 7.2.7***

For simplicity we take $S = \mathscr{O}$.

Let $\mathscr{U}$ be a finite open covering of $X$ by Stein manifolds $U_i$, for example we may take the $U_i$ to be coordinate neighbourhoods of unit discs. We can shrink $\mathscr{U}$ to a refinement $\mathscr{V} = \{V_j\}$ where each $V_j$ is a Stein manifold and each closure $\overline{V_j}$ is compact and contained in some $U_i$. Then all finite intersections $U_{i_0} \cap \cdots \cap U_{i_q}$, $V_{j_0} \cap \cdots \cap V_{j_q}$ are Stein manifolds, hence

$$H^p(U_{i_0} \cap \cdots \cap U_{i_q}, \mathscr{O}) = H^p(U_{j_o} \cap \cdots \cap V_{j_q}, \mathscr{O}) = 0$$

for all $p > 0$ by Theorem 7.2.6(2). By Theorem 7.2.8

$$H^q(X, \mathscr{O}) = H^q(\mathscr{U}, \mathscr{O}) = H^q(\mathscr{V}; \mathscr{O}) = 0 \text{ for all } q > 0.$$

Each $\mathscr{O}(U_{i_0} \cap \cdots \cap U_{i_q})$ is a Fréchet space. Since a finite product of Fréchet space is a Fréchet space, the product of all the vector spaces $\mathscr{O}(U_{i_0} \cap \cdots \cap U_{i_q})$ is a Fréchet space. $C^q(\mathscr{U}, \mathscr{O})$ is a closed subspace of the product, hence is a Fréchet space. The coboundary

$$\delta : C^q(\mathscr{U}, \mathscr{O}) \to C^{q+1}(\mathscr{U}, \mathscr{O})$$

is continuous, hence its kernel $Z^q(\mathscr{U}, \mathscr{O})$ is a Fréchet space.
Similarly with $\mathscr{V}$ in place of $\mathscr{U}$.
Hence $E = C^{q-1}(\mathscr{V}; \mathscr{O}) \times Z^q(\mathscr{U}, \mathscr{O})$ and $F = Z^q(\mathscr{U}, \mathscr{O})$ are Fréchet spaces.
Define $u, v : E \to F$ by

$$u(x, y) = \delta x + py, \qquad v(x, y) = -py$$

where $p : Z^q(\mathscr{U}, \mathscr{O}) \to Z^q(\mathscr{V}; \mathscr{O})$ is the restriction homomorphism from $U$ to $V$. If $w = u + v$, then $w(x, y) = \delta x$, and hence

$$\text{Coker } w = H^q(\mathscr{V}; \mathscr{O}) = H^q(X, \mathscr{O}).$$

By Theorem 7.2.10, the restriction $p$ is continuous, $u$ and $v$ are continuous.
Hence by Theorem 7.2.11,

$$\text{Coker } (u + v) = H^q(X, \mathscr{O})$$

is finite dimensional.
For an arbitrary coherent analytic sheaf $S$, the proof is essentially the same. □

TER

# 8 The de Rham Theorem

## 8.1 Exterior Product of Vector Space

Let $V$, $W$ be real vector spaces. $G$ the **free abelian group** generated by all pairs $(v, w) \in V \times W$, $H$ the subgroup of $G$ generated by all elements of the forms

$$(v_1 + v_2, w) - (v_1, w) - (v_2, w),$$
$$(v, w_1 + w_2) - (v, w_1) - (v, w_2),$$
$$(\lambda v, w) - (v, \lambda w)$$

where $v, v_1, v_2 \in V$, $w, w_1, w_2 \in W$, $\lambda \in \mathbb{R}$.

Then the factor group

$$G/H$$

is called the **tensor product** of $V$ and $W$, and is denoted by $V \otimes W$. The image of $(v, w)$ in $V \otimes W$ is denoted by $v \otimes w$.

If $\dim V = n$ and $\dim W = m$, and if $\{e_1, \cdots, e_n\}$ and $\{f_1, \cdots, f_m\}$ are bases of $V$ and $W$ respectively, then the $e_i \otimes f_i$, $i = 1, \cdots, n$, $j = 1, \cdots, m$ form a basis for $V \otimes W$, and thus $\dim V \otimes W = nm$.

If $U$ is another real vector space then we have a natural isomorphism, $(U \otimes V) \otimes W \simeq U \otimes (V \otimes W)$, and hence the expression $U \otimes V \otimes W$ is not ambiguous.

Let $V$ be a vector space. We can form its ***k-th*** **tensor product**

$$\otimes^k V = V \otimes \cdots \otimes V \ (k \text{ factors}).$$

Let $S_k$ be the symmetric group of $k$ elements $\{1, 2, \cdots, k\}$. Then $S_k$ acts by permutation of factors on $\otimes^k V$.

Let $T$ be the subspace of $\otimes^k V$ generated by all elements of the form

$$x - \mathrm{sign}(\sigma) \cdot \sigma(x), \; x \in \otimes^k V, \; \sigma \in S_k.$$

The quotient space

$$(\otimes^k V) / T$$

is called the ***k-th*** **exterior product** of $V$ and denoted by $\Lambda^k V$.

If $v_1, \cdots, v_k \in V$, then the image of $v_1 \otimes \cdots \otimes v_k$ in $\Lambda^k V$ is denoted by $v_1 \wedge \cdots \wedge v_k$. From the definition of $\Lambda^k V$, $v_1 \wedge \cdots \wedge v_k$ is a skew symmetric function of $v_1, \cdots, v_k$. For each integer $k \geq 0$, $\Lambda^k$ is a functor from the category of real vector spaces to itself, i.e., if $f: V \to W$ is a linear map, then $\Lambda^k f: \Lambda^k V \to \Lambda^k W$ and

$$\Lambda^k(f \circ g) = (\Lambda^k f) \circ (\Lambda^k g),$$
$$\Lambda^k(Id) = Id.$$

If $\{e_1, \cdots, e_n\}$ is a basis of $V$, then the elements

$$e_{i_1} \wedge \cdots \wedge e_{i_k}, \; 1 \leq i_1 < i_2 < \cdots < i_k \leq n,$$

form a basis of $\Lambda^k V$, so that $\dim(\Lambda^k V) = {}_n C_k$ and $\Lambda^k V = 0$ if $k > n$.

Conventionally we put $\Lambda^0 V = \mathbb{R}$, $\Lambda^1 V = V$, $\Lambda^n V \simeq \mathbb{R}$ with its generator $e_1 \wedge \cdots \wedge e_n$.

If $h: V \to V$ is a linear map then

$$\Lambda^n(h): \Lambda^n V \to \Lambda^n V,$$

and

$$\Lambda^n(h) = \det(h) \; (\in \mathbb{R}) : \mathbb{R} \to \mathbb{R},$$

is a multiplication of real number $\det(h)$.

## 8.2 Differential Form

Let $X$ be a $C^{\infty}$ differential manifold of dimension $n$. At each point $x \in X$ there is a tangent space $T_x X$, which is a real vector space of dimension $n$. If $x_1, \cdots, x_n$ are local coordinates at $x$, then the operators

$$\frac{\partial}{\partial x_i}, \quad i=1, \cdots, n$$

form a basis of $T_x X$. Let $T_x^* X$ be the dual space of $T_x X$, i.e., the vector space of all linear functions $T_x X \to \mathbb{R}$. The differentials

$$dx_i, \quad i = 1, \cdots, n$$

form a basis of $T_x^* X$.

The elements of $\Lambda^k T_x^* X$ are of the form

(1) $$w = \Sigma\, a_{i_1 \cdots i_k} dx_{i_1} \wedge \cdots \wedge dx_{i_k}$$

summed over all $k$-tuples $(i_1, \cdots, i_k)$ such that $1 \le i_1 < \cdots < i_k \le n$, the $a_{i_1 \cdots i_k}$ are real numbers.

The form $w$ is called a **differential $k$-form**, or simple **$k$-form**, at $x$.

Let $U$ be an open coordinate subset of $X$, then a $k$-form $w$ on $U$ is given locally by an expression (1) where each coefficient $a_{i_1 \cdots i_k}$ is a differentiable function of $x$ on $U$.

Thus a 0-form is simply a differentiable function $a(x)$, a 1-form is an expression $a_1(x)dx_1 + \cdots + a_n(x)dx_n$, and so on.

If $w$ is the $k$-form given by (1) on $U$, then we define the **exterior derivative** of $w$ by

$$dw = \Sigma\, d(a_{i_1 \cdots i_k}) \wedge dx_{i_1} \wedge \cdots \wedge dx_{i_k}$$

where $da = \Sigma_{j=1}^{n} \frac{\partial a}{\partial x_j} dx_j$.

Thus $dw$ is a $(k+1)$-form. By the skew symmetry we have

$$d(dw) = 0,$$

for all forms $w$.

## Definition 8.2.1

If $dw = 0$, then $w$ is called a **closed $k$-form**.
If $w = dv$ for some $(k-1)$-form $v$, then $w$ is called an **exact $k$-form**.
Since $d^2 = 0$, every exact form is closed and hence we can form the group $H^k_{deR}(X, \mathbb{R})$ of closed $k$-forms modulo exact $k$-forms.
The group $H^k_{deR}(X, \mathbb{R})$ is called the **$k$-th de Rham cohomology group** of $X$.

### Example 8.2.2

Let $X$ be a differentiable manifold of real dimension $n$ and let $\Omega^p(X)$ be the sheaf of real-valued differential forms of degree $p$. Then there is a resolution of the constant sheaf $\mathbb{R}$ given by

$$0 \longrightarrow \mathbb{R} \xrightarrow{i} \Omega^0(X) \xrightarrow{d} \Omega^1(X) \xrightarrow{d} \cdots \longrightarrow \Omega^n(X) \longrightarrow 0,$$

where $i$ is the natural inclusion and $d$ is the exterior differentiation operator.
Since $d^2 = 0$, it is clear that the above is a differential sheaf. However, the classical Poincaré lemma (Theorem 8.3.4) asserts that on a star-shaped domain $U$ in $\mathbb{R}^n$, if $f \in \Omega^p(U)$ is given such that $df = 0$, then there exists a $u \in \Omega^{p-1}(U)$ $(p > 0)$ so that $du = f$. Therefore the induced mapping $d_x$ on the stalks at $x \in X$ is exact, since we can find

representatives in local coordinates in star-shaped domains. At the term $\Omega^0(X)$, exactness is an elementary result from calculus [i.e., $df \equiv 0$ implies that $f$ is a constant(locally)]. We shall denote this resolution by $0 \to \mathbb{R} \to \Omega^*(X)$ (or $0 \to \mathbb{C} \to \Omega^*(X)$ if we are using complex coefficients).

## 8.3 The de Rham Theorem for Differential Manifold

Let $X$ be a differentiable manifold. For each integer $k \geq 0$ let $S_k(X)$ be the real vector space generated by the differentiable singular $k$-simplecies in $X$. The element of $S_k(X)$ is a finite linear combination of differentiable singular $k$-simplecies with real coefficients which is called a differentiable singular $k$-chain in $X$. For $k < 0$ let $S_k(X)$ be the zero vector space. The boundary operator $\partial$ defined by $\partial\sigma = \sum_{i=0}^{k}(-1)^i\sigma^i$ for a singular $k$-simplex $\sigma$, induces linear transformations

$$\partial_k : S_k(X) \to S_{k-1}(X)$$

for each integer $k$, which for $k < 0$ are the zero transformation. By simple calculation, for all $k$

$$\partial_{k-1} \circ \partial_k = 0.$$

The **$k$-th differential singular homology group of $X$ with real coefficients** is defined by

$$H_k(X, \mathbb{R}) = \mathrm{Ker}\,\partial_k / \mathrm{Im}\,\partial_{k+1},$$

which is a real vector space.

The elements of $\mathrm{Ker}(\partial_k)$ are called **differentiable $k$-cycles**, and the elements of $\mathrm{Im}(\partial_{k+1})$ are called **differentiable $k$-boundaries**.

Let $X$ be a connected paracompact differentiable manifold of dimension $n$, $C$ a $k$-dimensional chain on $X$, $w$ a $k$-form defined in some open neighbourhood of $C$ in $X$.

Then we can integrate $w$ on the chain $C$ and get a real number $\int_C w$.

**Theorem 8.3.1 Stokes**

Let $C$ be a $(k+1)$-chain on $X$, $\partial C$ its boundary, $w$ a $k$-form defined in a neighbourhood of $C$. Then

$$\int_C dw = \int_{\partial C} w.$$

***Remark 8.3.2***

If $w$ is a closed $k$-form and $C$ a $k$-cycle with real coefficients then $w$ gives rise to a homomorphism $C \mapsto \int_C w := <C, w>$, of $Z^k(X, \mathbb{R})$ into $\mathbb{R}$.

From the Stokes' Theorem,

(1) the real number $<C, w>$ depends only on the class $w$ in the de Rham group $H^k_{deR}(X, \mathbb{R})$, and the class $C$ in the singular homology group $H_k(X, \mathbb{R})$

(2) $<\ ,\ w>$ maps the $k$-boundaries to 0 and hence induces a homomorphism

$$<\ ,\ w>:\ H_k(X, \mathbb{R}) \to \mathbb{R},$$

as an element of $H_k(X, \mathbb{R})^* = H^k(X, \mathbb{R})$.

Hence we have a homomorphism

$$<\ ,\ >:\ H^k_{deR}(X, \mathbb{R}) \to H^k(X, \mathbb{R}),\ w \mapsto <\ ,\ w>$$

for each $w \in H^k_{deR}(X, \mathbb{R})$.

**Theorem 8.3.3** **de Rham**

$<\ ,\ >: H^k_{deR}(X, \mathbb{R}) \to H^k(X, \mathbb{R})$ is an isomorphism for $k=0, 1, \cdots, n$.

We shall show that $H^k_{deR}(X, \mathbb{R}) \cong H^k(X, \mathbb{R})$ by using the sheaf theory.

Let $\Omega^k(X)$ denote the sheaf of germs of differentiable $k$-forms on $X$, $k=0, \cdots, n$, thus $\Omega^0(X) = \mathscr{E}$ is the **sheaf of germs of differentiable functions on** $X$. We have a sheaf homomorphism

$$d : \Omega^k(X) \to \Omega^{k+1}(X).$$

Let $\Phi^k(X) :=$ the kernel of $d$ be the sheaf of germs of closed $k$-forms.

**Theorem 8.3.4** **Poincaré Lemma**

If $w$ is a closed $k$-form in the open ball $U = \{x \in \mathbb{R}^n \mid |x| < 1\}$ then there is a differentiable $(k-1)$-form $U$ defined in some open ball $|x| < \eta$ $(0 < \eta \leq 1)$ such that $w = dU$ (where $k \geq 1$).
That is, $H^k(U, \mathbb{R}) = 0$ for all $k > 0$.

***Proof.*** Construct linear maps

$$h_k : \Omega^{k+1}(U) \to \Omega^k(U)$$

such that $h_k \circ d + d \circ h_{k-1} =$ identity

This is done through an integration process. Since these maps are to be linear, it suffices to define $h_{k-1}$ on forms $\omega = g dx_{i_1} \wedge \cdots \wedge dx_{i_k}$; similarly for $h_k$.

For such $\omega$, set

$$h_{k-1}(\omega)(x) = \left(\int_0^1 t^{k-1} g(tx) dt\right)\mu,$$

where

$$\mu = x_{i_1} dx_{i_2} \wedge \cdots \wedge dx_{i_k} - x_{i_2} dx_{i_1} \wedge dx_{i_3} \wedge \cdots \wedge dx_{i_k}$$
$$+ \cdots + (-1)^{k-1} x_{i_k} dx_{i_1} \wedge \cdots \wedge dx_{i_{k-1}}.$$

Then $d\mu = k\, dx_{i_1} \wedge \cdots \wedge dx_{i_k}$.

The map $h_k$ is defined similarly by replacing $k$ everywhere by $k+1$.

Now, for $\omega = g\, dx_{i_1} \wedge \cdots \wedge dx_{i_k} \in \Omega^k(U)$ and $x \in U$,

$$\begin{aligned}
&(d \circ h_{k-1})(\omega)(x) \\
&= d[(\int_0^1 t^{k-1} g(tx)\, dt)\mu] \\
&= \sum_{j=1}^{n} \frac{\partial}{\partial x_j}\left(\int_0^1 t^{k-1} g(tx)\, dt\right) dx_j \wedge \mu + \left(\int_0^1 t^{k-1} g(tx)\, dt\right) d\mu \\
&= \sum_{j=1}^{n} \left(\int_0^1 t^{k-1} \frac{\partial}{\partial x_j}(g(tx)) dt\right) dx_j \wedge \mu + \left(\int_0^1 t^{k-1} g(tx)\, dt\right) d\mu \\
&= \sum_{j=1}^{n} \left(\int_0^1 t^{k} \frac{\partial g}{\partial x_j}(tx) dt\right) dx_j \wedge \mu + k\left(\int_0^1 t^{k-1} g(tx)\, dt\right) dx_{i_1} \wedge \cdots \wedge dx_{i_k},
\end{aligned}$$

and

$$\begin{aligned}
(h_k \circ d)(\omega)(x) &= h_k\left(\sum_{j=1}^{n} \frac{\partial g}{\partial x_j} dx_j \wedge dx_{i_1} \wedge \cdots \wedge dx_{i_k}\right) \\
&= \sum_{j=1}^{n} \left(\int_0^1 t^k \frac{\partial g}{\partial x_j}(tx) dt\right) [x_j dx_{i_1} \wedge \cdots \wedge dx_{i_k} - dx_j \wedge \mu].
\end{aligned}$$

Thus,

$$\begin{aligned}
&(d \circ h_{k-1} + h_k \circ d)(\omega)(x) \\
&= \left[k\left(\int_0^1 t^{k-1} g(tx) dt\right) + \sum_{j=1}^{n}\left(\int_0^1 t^k \frac{\partial g}{\partial x_j}(tx) x_j\, dt\right)\right] dx_{i_1} \wedge \cdots \wedge dx_{i_k} \\
&= \left\{\int_0^1 \left[k t^{k-1} g(tx) + t^k \frac{d}{dt}(g(tx))\right] dt\right\} dx_{i_1} \wedge \cdots \wedge dx_{i_k}
\end{aligned}$$

$$= \left\{ \int_0^1 \frac{d}{dt} [t^k g(tx)] dt \right\} dx_{i_1} \wedge \cdots \wedge dx_{i_k}$$

$$= t^k g(tx) \big|_0^1 dx_{i_1} \wedge \cdots \wedge dx_{i_k}$$

$$= g(x) dx_{i_1} \wedge \cdots \wedge dx_{i_k}$$

$$= \omega(x) \text{ (for all } x \in U).$$

Suppose $\omega \in \Omega^k(U)$ is closed, then

$$\omega = (h_k \circ d + d \circ h_{k-1})(\omega) = h_k(d\omega) + d(h_{k-1}\omega)$$
$$= d(h_{k-1}\omega).$$

Thus $H^k(U, \mathbb{R}) = 0$ for all $k > 0$. □

### *Remark 8.3.5*

In $\mathbb{R}^3$ the sequence of sheaves

$$0 \to \mathbb{R} \to \Omega^0(\mathbb{R}^3) \to \Omega^1(\mathbb{R}^3) \to \Omega^3(\mathbb{R}^3) \to 0$$

is exact.

If $f \in \Omega^0(\mathbb{R}^3)$, $f : \mathbb{R}^3 \to \mathbb{R}$, then

$$df = \frac{\partial f}{\partial x_1} dx_1 + \frac{\partial f}{\partial x_2} dx_2 + \frac{\partial f}{\partial x_3} dx_3$$
$$= \left( \frac{\partial f}{\partial x_1}, \frac{\partial f}{\partial x_2}, \frac{\partial f}{\partial x_3} \right) := \textbf{gradient}(f).$$

Thus $df = 0$ if and only if $f$ is a constant.

If $\omega = f_1 dx_1 + f_2 dx_2 + f_3 dx_3$, $k^1(\mathbb{R}^3)$, then

$$d\omega = \left( \frac{\partial f_3}{\partial x_2} - \frac{\partial f_2}{\partial x_3} \right) dx_2 \wedge dx_3 + \left( \frac{\partial f_1}{\partial x_3} - \frac{\partial f_3}{\partial x_1} \right) dx_3 \wedge dx_1$$
$$+ \left( \frac{\partial f_2}{\partial x_1} - \frac{\partial f_1}{\partial x_2} \right) dx_1 \wedge dx_2.$$

The **curl** $(f_1, f_2, f_3) := \left(\frac{\partial f_3}{\partial x_2} - \frac{\partial f_2}{\partial x_3}, \frac{\partial f_1}{\partial x_3} - \frac{\partial f_3}{\partial x_1}, \frac{\partial f_2}{\partial x_1} - \frac{\partial f_1}{\partial x_2}\right).$

If $\omega = f_1 dx_2 \wedge dx_3 + f_2 dx_3 \wedge dx_1 + f_3 dx_1 \wedge dx_2 \in \Omega^2(\mathbb{R}^3)$, then

$$\begin{aligned} d\omega &= \left(\frac{\partial f_1}{\partial x_1} + \frac{\partial f_2}{\partial x_2} + \frac{\partial f_3}{\partial x_3}\right) dx_1 \wedge dx_2 \wedge dx_3 \\ &:= \mathbf{div}\,(f_1, f_2, f_3) dx_1 \wedge dx_2 \wedge dx_3. \end{aligned}$$

We have the famous formulas:

curl $\circ$ gradient $= 0$,

divergence $\circ$ curl $= 0$.

Recall that $\Phi^k(X) =$ the kernel $(d : \Omega^k(X) \to \Omega^{k+1}(X))$ is the sheaf of germs of closed $k$-forms on $X$.

From the Poincaré Theorem 8.3.4 we have an exact sequence of sheaves

$$0 \to \Phi^k(X) \to \Omega^k(X) \to \Phi^{k+1}(X) \to 0.$$

Since exactness is a local property, and hence we have an exact cohomology sequence

$$\begin{aligned} (*) \quad \cdots &\to H^{p-1}(X, \Omega^k(X)) \to H^{p-1}(X, \Phi^{k+1}(X)) \\ &\to H^p(X, \Phi^k(X)) \to \cdots. \end{aligned}$$

Recall that $\mathscr{E}$ is the sheaf of germs of differentiable functions on $X$.

**Theorem 8.3.6**

If $S$ is a sheaf of $\mathscr{E}$-modules on $X$, then

$$H^q(X, S) = 0$$

for all $q > 0$.

***Proof.*** We shall use the Čech cohomology since $X$ is paracompact.

Let $\mathscr{U} = \{U_i\}$ be an open covering of $X$. We may assume that $\mathscr{U}$ is locally finite.

By paracompactness there is a **partition of unity differentiable functions** $\{\phi_i : U_i \to \mathbb{R}\}$ and closed subsets $V_i \subset U_i$ such that

(1) the $V_i$ cover $X$

(2) $\phi_i \geq 0$ on $U_i$ and $\phi_i \equiv 0$ on $U_i - V_i$

(3) for each $x \in X$, $\Sigma_i \phi_i(x) = 1$.

We shall prove that $H^1(X, S) = 0$.

Let $f = \{f_{ij}\}$ be a 1-cocycle of $\mathscr{U}$ with values in $S$. Define

$$g_i = \Sigma_k \phi_k f_{ki}.$$

Since $S$ is a sheaf of $\mathscr{E}$-modules and $\phi_k \in \mathscr{E}(U_k)$, we have $g_i \in S(U_i)$, and hence $g$ is a 1-cochain of $\mathscr{U}$ with values in $S$.

$$\begin{aligned}(\delta g)_{ij} = g_j - g_i &= \Sigma_k \phi_k (f_{kj} - f_{ki}) \\ &= \Sigma_k \phi_k f_{ij} = f_{ij}\end{aligned}$$

since $f_{ij} - f_{kj} + f_{ki} = 0$ and $\Sigma \phi_k = 1$.

Hence the 1-cocycle $f$ is a 1-coboundary.

Hence $H^1(\mathscr{U}, S) = 0$ and $H^1(X, S) = 0$.

For $p > 1$ the proof is analogous. □

***Proof of 8.3.3.***

Since each $\Omega^k(X)$ is a sheaf of $\mathscr{E}$-modules, by (8.3.6) we have $H^p(X, \Omega^k(X)) = 0$ for all $p > 0$.

From the exact sequence $(*)$ we have isomorphisms

$$H^{p-1}(X, \Phi^{k+1}(X)) \cong H^p(X, \Phi^k(X)) \text{ for } p > 1 \text{ and } k \geq 0.$$

Let $q = p + k$. We have a chain of isomorphisms;

$$H^1(X, \Phi^{q-1}(X)) \cong H^2(X, \Phi^{q-2}(X))$$
$$\cdots$$
$$\cong H^q(X, \Phi^0(X)) = H^q(X, \mathbb{R})$$

where $\Phi^0(X)$ is the constant sheaf of real numbers.

If we consider the left hand end of the exact sequence $(*)$, we have an exact sequence

$$0 \to H^0(X, \Phi^{q-1}(X)) \to H^0(X, \Omega^{q-1}(X)) \to H^0(X, \Phi^q(X))$$
$$\to H^1(X, \Phi^{q-1}(X)) \to 0.$$

Hence $H^1(X, \Phi^{q-1}(X)) \cong H^0(X, \Omega^q(X))/\mathrm{Im}\, H^0(X, \Phi^{q-1}(X))$

$\cong H^q_{deR}(X, \mathbb{R})$ : a de Rham cohomology group.

Hence we have

$$H^q_{deR}(X, \mathbb{R}) \cong H^q(X, \mathbb{R}) \text{ for } q > 0.$$

If $q = 0$, $H^0_{deR}(X, \mathbb{R}) = \mathbb{R}$, $H_0(X, \mathbb{R})^* = \mathbb{R}$ since $X$ is connected. Thus

$$H^q_{deR}(X, \mathbb{R}) \cong H^q(X, \mathbb{R}) \text{ for all } q \geq 0. \qquad \square$$

## 8.4 The de Rham Theorem for Stein Manifold

Let $X$ be a connected complex manifold of complex dimension $n$. Let us consider **holomorphic $k$-forms** instead of differential $k$-forms;

$$w = \sum a_{i_1 \cdots i_k} dz_{i_1} \wedge \cdots \wedge dz_{i_k}$$

where the coefficients $a_{i_1 \cdots i_k}$ are holomorphic functions of the local coordinates $z_1, \cdots, z_n$.

As before we have a sheaf $\Omega^k(X)$ of **germs of holomorphic $k$-forms**, and a sheaf holomorphism

$$d: \Omega^k(X) \to \Omega^{k+1}(X),$$

and hence we have **de Rham cohomology groups** $H^k_{deR}(X, \mathbb{C})$

**Theorem 8.4.1 de Rham Theorem for Stein Manifold**

Let $X$ be a connected Stein manifold of dimension $n$.
Then

$$H^k_{deR}(X, \mathbb{C}) \cong H^k(X, \mathbb{C}) \quad \text{for} \quad k = 0, 1, \cdots, n.$$

***Proof.*** The Poincaré Lemma is still true in the holomorphic case.

Recall Theorem 7.2.6(2) of Chapter 7:

If $X$ is a Stein manifold and $S$ is a coherent analytic sheaf on $X$, then

$$H^q(X, S) = 0 \quad \text{for all } q > 0.$$

Since $\Omega^k$ is a coherent sheaf of $\mathcal{O}$-modules, which is locally isomorphic to $\mathcal{O}^k$, $H^p(X, \Omega^k(X)) = 0$ for all $p > 0$.

Let $\Phi^k(X) = \operatorname{Ker} d: \Omega^k(X) \to \Omega^{k+1}(X)$. By Poincaré Theorem, we have an exact sequence of sheaves

$$0 \to \Phi^k(X) \to \Omega^k(X) \to \Phi^{k+1}(X) \to 0$$

and hence an exact cohomology sequence

$$\cdots \to H^{p-1}(X, \Phi^k(X)) \to H^{p-1}(X, \Phi^{k+1}(X)) \to H^p(X, \Phi^k(X))$$
$$\to H^p(X, \Omega^k(X)) \to \cdots.$$

We may use Theorem 7.2.6(2) of Chapter 7 to prove this theorem as before. □

**Corollary 8.4.2** $H^k(X, \mathbb{C}) = 0$ for $k > n$.

Since there are no holomorphic $k$-forms if $k > n$.

## 8.5 Lefschetz Theorem

Let $X \subset \mathbb{CP}^m$ be an irreducible algebraic submanifold of $\mathbb{CP}^m$ of dimension $n$. Choose a hyperplane $H \subset \mathbb{CP}^m$ which cuts $X$ transversally, i.e., for each $y \in Y = X \cap H$ the projective tangent space of $X$ at $y$ is not contained in $H$. Then $Y = X \cap H$ has dimension $n-1$.

**Theorem 8.5.1 Lefschetz**

The inclusion $i : Y \to X$ induces isomorphisms

$$i_* : H_q(Y, \mathbb{C}) \to H_q(X, \mathbb{C})$$

for $q = 0, 1, \cdots, n-2$, and an epimorphism

$$i_* : H_{n-1}(Y, \mathbb{C}) \to H_{n-1}(X, \mathbb{C}).$$

***Proof.*** $\mathbb{CP}^m - H \cong \mathbb{C}^m$ is an affine space of complex dimension $m$.

$X - Y$ is a closed complex submanifold of $\mathbb{C}^m$.

Hence $X - Y$ is a Stein manifold. By Corollary 8.4.2,

$$H^q(X - Y, \mathbb{C}) = 0 \quad \text{for } q > n.$$

By the Alexander Duality Theorem for the pair $(X, Y)$,

$$H^q(X - Y, \mathbb{C}) \cong H_{2n-q}(X, Y; \mathbb{C}).$$

From the exact homology sequence for the pair $(X, Y)$

$$\cdots \to H_{q+1}(X, Y; \mathbb{C}) \to H_q(Y, \mathbb{C}) \to H_q(X, \mathbb{C})$$

$$\to H_q(X,\ Y;\ \mathbb{C}) \to \cdots .$$

Since $H_0(X,\ Y;\ \mathbb{C}) = H_1(X,\ Y;\ \mathbb{C}) = \cdots = H_{n-1}(X,\ Y;\ \mathbb{C}) = 0$, we have the result. □

***Remark 8.5.2***

(1) When $q = 0$, Theorem 8.5.1 tells us that if $X$ is connected and $n > 1$, then $Y$ is connected.
When $q = 1$, $n > 1$, that every 1-cycle in $X$ is homologous in $X$ to a 1-cycle in $Y$.

(2) In general $H_{n-1}(Y,\ \mathbb{C}) \to H_{n-1}(X,\ \mathbb{C})$ is not an isomorphism.
For example, let $X$ be a cubic surface in $\mathbb{CP}^3$. Then $X$ is homeomorphic to $\mathbb{CP}^2 \# 6\overline{\mathbb{CP}^2}$, and $Y$ is a non-singular cubic curve which is a Riemann surface of genus 1, see Example 2 of Appendix A. Therefore we have $H_1(X,\ \mathbb{C}) = 0$, but $H_1(Y,\ \mathbb{C}) = \mathbb{C}^2$.

(3) Under the hypothesis of Theorem 8.5.1, the relative homotopy group

$$\pi_k(X,\ Y) = 0 \text{ for } k < n.$$

□

TER

9

# Hodge Theorem

## 9.1 Laplace Operator

Let $X$ be a compact oriented Riemannian manifold of dimension $n$. We will see that the ordinary Laplacian $(-1)\sum_i \partial^2/\partial x_i^2$ has a generalization to an operator $\triangle$ on differential forms, known as the Laplace operator. Our main objective in this chapter is a proof of the Hodge decomposition theorem, which says that the equation $\triangle\omega = \alpha$ has a solution $\omega$ in the smooth $p$-forms on $X$ if and only if the $p$-form $\alpha$ is orthogonal (in a suitable inner product on $\Omega^p(X)$) to the space of harmonic $p$-forms (those for which $\triangle\varphi = 0$). From the Hodge decomposition theorem we will conclude that there exists a unique harmonic form in each de Rham cohomology class.

Let $X$ be a compact oriented Riemannian manifold of dimension $n$. Recall the **Hodge $*$-operator**

$$* : \wedge^p T_x^* X \to \wedge^{n-p} T_x^* X$$

is given by the orientation and the inner product at $x$ for each $\alpha, \beta \in \wedge^p T_x^* X$,

$$\alpha \wedge *\beta = <\alpha, \beta> \text{ vol},$$

where vol is volume form at $x \in X$.

Hereafter $\Omega^p(X)$ is the set of all smooth $p$-forms on $X$.

Then the linear operator $*$ which assigns to each $p$-form on $X$ an $(n-p)$-form and which satisfies

$$** = (-1)^{p(n-p)}. \tag{1}$$

We define a **dual operator** $\delta$ from $p$-forms to $(p-1)$-forms by setting

$$\delta = (-1)^{n(p+1)+1} * d *. \tag{2}$$

On 0-forms, $\delta$ is simply the zero linear functional.

**The Laplace operator** $\triangle$ is defined by

$$\triangle = \delta d + d\delta, \tag{3}$$

and is a linear operator on $\Omega^p(X)$ for each $p$ with $0 \leq p \leq n.$

For $C^\infty$ functions on Euclidean space $\mathbb{R}^n$, the Laplacian is simply the operator

$$\triangle = (-1)\sum_{i=1}^{n} \partial^2/\partial x_i^2.$$

Also, the Laplacian commutes with $*$, that is,

$$* \triangle = \triangle *. \tag{4}$$

We define an **inner product** on the vector space $\Omega^p(X)$ of $p$-forms on $X$ by setting

$$< \alpha, \beta >= \int_M \alpha \wedge * \beta \text{ for } \alpha, \beta \in \Omega^p(X), \tag{5}$$

and we denote the corresponding norm of $\alpha$ by $\|\alpha\|$. The bilinear form defined in (5) is actually symmetric and positive definite. We extend the inner products (5) for $0 \leq p \leq n$ to an inner product on the direct sum $\bigoplus_{p=0}^{n} \Omega^p(X)$ simply by requiring the various $\Omega^p(X)$ to be orthogonal.

**Theorem 9.1.1**

The $\delta$ is the adjoint of the exterior derivative $d$ on $\bigoplus_{p=0}^{n} \Omega^p(X)$, that is,

$$< d\alpha, \beta >=< \alpha, \delta\beta >.$$

***Proof.*** By linearity, and the orthogonality of the $\Omega^p(X)$, the proof reduces to consideration of the case in which $\alpha$ is a $(p-1)$-form and $\beta$ is a $p$-form. In this case,

$$\begin{aligned} d(\alpha \wedge * \beta) &= d\alpha \wedge * \beta + (-1)^{p-1} \alpha \wedge d * \beta \\ &= d\alpha \wedge * \beta - \alpha \wedge * \delta\beta. \end{aligned}$$

By integrating both sides over $X$ and applying Stokes' Theorem to the left-hand side, we obtain

$$0 = \int_X (d\alpha \wedge * \beta - \alpha \wedge * \delta\beta) = < d\alpha, \beta > - < \alpha, \delta\beta >.$$

Thus

$$< d\alpha, \beta > = < \alpha, \delta\beta >.$$ □

**Theorem 9.1.2**

The Laplace operator $\Delta$ is self-adjoint, that is,

$$< \Delta\alpha, \beta > = < \alpha, \Delta\beta > \quad \alpha, \beta \in \Omega^p(X),\ 0 \le p \le n.$$

**Theorem 9.1.3**

$\Delta\alpha = 0$ if and only if $d\alpha = 0$ and $\delta\alpha = 0$.

***Proof.*** Clearly $\Delta\alpha = 0$ if $d\alpha = 0$ and $\delta\alpha = 0$. Now,

$$< \Delta\alpha, \alpha > = < (d\delta + \delta d)\alpha, \alpha > = < \delta\alpha, \delta\alpha > + < d\alpha, d\alpha >.$$

Thus if $\Delta\alpha = 0$, it follows that $d\alpha = 0$ and $\delta\alpha = 0$. □

**Theorem 9.1.4**

The only harmonic functions $(\Delta f = 0)$ on a compact, connected, oriented, Riemannian manifold are the constant functions.

## 9.2 Hodge Decomposition Theorem on Compact Oriented Manifold

We shall let $\triangle^*$ denote the adjoint of the Laplacian on $\Omega^p(X)$. This operator is, of course, precisely $\triangle$ itself since the Laplacian is self-adjoint on $\Omega^p(X)$, and usually we make no distinction between $\triangle$ and $\triangle^*$. However, this distinction will be important for the form of the following definition.

We are interested in finding necessary and sufficient conditions to exist a solution $\omega$ of the equation $\triangle\omega=\alpha$. Suppose that $\omega$ is a solution of $\triangle\omega=\alpha$. Then

$$<\triangle\omega,\ \varphi>=<\alpha,\ \varphi> \text{ for all } \varphi\in\Omega^p(X), \tag{1}$$

from which it follows that

$$<\omega,\ \triangle^*\varphi>=<\alpha,\ \varphi> \text{ for all } \varphi\in\Omega^p(X). \tag{2}$$

Now, (2) suggests that we can view a solution of $\triangle\omega=\alpha$ as a certain type of a linear function on $\Omega^p(X)$, namely, determines a bounded linear function $l$ on $\Omega^p(X)$ by

$$l(\beta)=<\omega,\ \beta>; \tag{3}$$

and in view of $l(2)$, the function $l$ satisfies

$$l(\triangle^*\varphi)=<\alpha,\ \varphi> \text{ for all } \varphi\in\Omega^p(X).$$

This view of a solution turns out to be extremely useful, for it will allow us to bring various techniques of function analysis to bear on the problem of solving $\triangle\omega=\alpha$. We shall call such a linear function a **weak solution** of $\triangle\omega=\alpha$. That is, a weak solution of $\triangle\omega=\alpha$ is a bounded linear function $l:\Omega^p(X)\rightarrow\mathbb{R}$ such that

$$l(\Delta^* \varphi) = < \alpha, \varphi > \text{ for all } \varphi \in \Omega^p(X). \tag{5}$$

We have seen that each ordinary solution $\omega \in \Omega^p(X)$ of $\Delta\omega = \alpha$ determines a weak solution by (3). It turns out that the major effort of this chapter will be to prove a regularity theorem which says that the converse of this is true; that is, each weak solution determines an ordinary solution.

The main step in proving this converse is to show that if $l$ is a weak solution of $\Delta\omega = \alpha$, then $l$ is represented by a smooth form $\omega$ in the sense that there is a form $\omega \in \Omega^p(X)$ such that (3) holds. That the form $\omega$ is then an ordinary solution follows from the fact that

$$\begin{aligned} < \Delta\omega, \beta > &= < \omega, \Delta^* \beta > \\ &= l(\Delta^* \beta) = < \alpha, \beta > \end{aligned}$$

for all $\beta \in \Omega^p(X)$, which implies that $\Delta\omega = \alpha$.

**Theorem 9.2.1**

Let $\alpha \in \Omega^p(X)$, and let $l$ be a weak solution of $\Delta\omega = \alpha$. Then there exists $\omega \in \Omega^p(X)$ such that

$$l(\beta) = < \omega, \beta >$$

for every $\beta \in \Omega^p(X)$. Consequently, $\Delta\omega = \alpha$.

We shall assume Theorem 9.2.1 and Theorem 9.2.2.

**Theorem 9.2.2**

Let $\{\alpha_n\}$ be a sequence of smooth $p$-forms on $X$ such that $\|\alpha_n\| \le c$ and $\|\Delta\alpha_n\| \le c$ for all $n$ and for some constant $c > 0$. Then a subsequence of $\{\alpha_n\}$ is a Cauchy sequence in $\Omega^p(X)$.

**Definition 9.2.3**

We let

$$H^p(X) := \{\omega \in \Omega^p(X) : \triangle\omega = 0\}.$$

The elements of $H^p(X)$ are called **harmonic $p$-forms.**

**Theorem 9.2.4 Hodge**

Let $X$ be a compact oriented Riemannian manifold of dimension $n$. For each integer $p$ with $0 \leq p \leq n$, $H^p(X)$ is finite dimensional, and we have the following orthogonal direct sum decompositions of the space $\Omega^p(X)$ of smooth $p$-forms on $X$:

$$\begin{aligned}\Omega^p(X) &= \triangle(\Omega^p)(X) \oplus H^p(X) \\ &= d\delta(\Omega^p(X)) \oplus \delta d(\Omega^p(X)) \oplus H^p(X) \\ &= d(\Omega^{p-1}(X)) \oplus \delta(\Omega^{p+1}(X)) \oplus H^p(X) \qquad (1)\end{aligned}$$

Consequently, the equation $\triangle\omega = \alpha$ has a solution $\omega = \Omega^p(X)$ if and only if the $p$-form $\alpha$ is orthogonal to the space of harmonic $p$-forms.

***Proof.*** If $H^p(X)$ were not finite dimensional, then $H^p(X)$ would contain an orthonormal sequence. But by Theorem 9.2.2, this orthonormal sequence would contain a Cauchy subsequence, which is impossible. Thus $H^p(X)$ is finite dimensional.

It is sufficient to prove the decomposition in the first line of (1) for the other two lines of (1) then follow from $\triangle = \delta d + d\delta$, (9.1.1) and (9.1.3)

Let $\omega_1, \cdots, \omega_l$ be an orthonormal basis of $H^p(X)$. Then an arbitrary form $\alpha \in \Omega^p(X)$ can uniquely be written

$$\alpha = \beta + \sum_{i=1}^{l} < \alpha, \ \omega_i > \omega_i \tag{2}$$

where $\beta$ lies in $(H^p(X))^{\perp}$, the subspace of $\Omega^p(X)$ consisting of all elements orthogonal to $H^p(X)$. Thus we have an orthogonal direct sum decomposition

$$\Omega^p(X) = (H^p(X))^{\perp} \oplus H^p(X). \tag{3}$$

The theorem will be proved by showing that $(H^p(X))^{\perp} = \triangle(\Omega^p(X))$. We let $H$ denote the **projection operator** of $\Omega^p(X)$ onto $H^p(X)$ so that $H(\alpha)$ is the harmonic part of $\alpha$.
Now $\triangle(\Omega^p(X)) \subset (H^p(X))^{\perp}$. For if $\omega \in \Omega^p(X)$ and $\alpha \in H^p(X)$, then

$$< \triangle\omega, \ \alpha >=< \omega, \ \triangle\alpha >= 0.$$

Conversely, we claim that $(H^p(X))^{\perp} \subset \triangle(\Omega^p(X))$. In order to prove this, we first need the following inequality.
We claim that there is a constant $c > 0$ such that

$$\|\beta\| \leq c\|\triangle\beta\| \text{ for all } \beta \in (H^p(X))^{\perp}. \tag{4}$$

Suppose the contrary. Then there exists a sequence $\beta_j \in (H^p(X))^{\perp}$ with $\|\beta_j\| = 1$ and $\|\triangle\beta_j\| \to 0$. By Theorem 9.2.2 a subsequence of the $\beta_j$, which for convenience we can assume to be $\{\beta_j\}$ itself, is Cauchy. Thus $\lim < \beta_j, \ \psi >$ exists for each $\psi \in \Omega^p(X)$. We define a linear functional $l$ on $\Omega^p(X)$ by setting

$$l(\psi) = \lim_{j \to \infty} < \beta_j, \ \psi > \text{ for } \psi \in \Omega^p(X).$$

Now $l$ is clearly bounded, and

$$l(\triangle\psi) = \lim_{j \to \infty} < \beta_j, \ \triangle\psi >= \lim_{j \to \infty} < \triangle\beta_j, \ \psi >= 0,$$

so $l$ is a weak solution of $\Delta\beta = 0$. By Theorem 9.2.1 there exists $\beta \in \Omega^p(X)$ such that $l(\psi) = <\beta, \psi>$. Consequently, $\beta_j \to \beta$. Since $\|\beta_j\| = 1$ and $\beta_j \in (H^p(X))^\perp$, it follows that $\|\beta\| = 1$ and $\beta \in (H^p(X))^\perp$. But by Theorem 9.2.1, $\Delta\beta = 0$, so $\beta \in H^p(X)$, which is a contradiction. Thus (4) is proved.

Now we shall use (4) to prove that $(H^p(X))^\perp \subset \Delta(\Omega^p(X))$. Let $\alpha \in (H^p(X))^\perp$.

We define a linear functional $l$ on $\Delta(\Omega^p(X))$ by setting

$$l(\Delta\varphi) = <\alpha, \varphi> \text{ for all } \varphi \in \Omega^p(X).$$

Now $l$ is well-defined; for if $\Delta\varphi_1 = \Delta\varphi_2$, then $\varphi_1 - \varphi_2 \in H^p(X)$, so that $<\alpha, \varphi_1 - \varphi_2> = 0$. Also $l$ is a bounded linear functional on $\Delta(\Omega^p(X))$, for let $\varphi \in \Omega^p(X)$ and let $\psi = \varphi - H(\varphi)$. Then by using (4), we obtain

$$\begin{aligned} |l(\Delta\varphi)| = |l(\Delta\psi)| &= |<\alpha, \psi>| \leq \|\alpha\|\|\psi\| \\ &\leq c\|\alpha\|\|\Delta\psi\| = c\|\alpha\|\|\Delta\varphi\|. \end{aligned}$$

By the Hahn-Banach theorem, $l$ extends to a bounded linear functional on $\Omega^p(X)$. Thus $l$ is a weak solution of $\Delta\omega = \alpha$.

By Theorem 9.2.1, there exists $\omega \in \Omega^p(X)$ such that $\Delta\omega = \alpha$. Hence

$$(H^p(X))^\perp = \Delta(\Omega^p(X)),$$

and the Hodge theorem is proved. □

We define the **Green operator** $G : \Omega^p(X) \to (H^p(X))^\perp$ by setting $G(\alpha)$ equal to the unique solution of $\Delta\omega = \alpha - H(\alpha)$ in $(H^p(X))^\perp$.

**Theorem 9.2.5**

The Green operator $G$ commutes with $d$, $\delta$, and $\triangle$. Moreover $G$ commutes with any linear operator which commutes with the Laplacian $\triangle$.

***Proof.*** Suppose that $T\triangle = \triangle T$ with, say, $T: \Omega^p(X) \to \Omega^q(X)$. Let $\pi$ denote the projection mapping of $\Omega^p(X)$ onto $(H^p(X))^\perp$. By definition, $G = (\triangle | (H^p(X))^\perp)^{-1} \circ \pi$. Now, the fact that $T\triangle = \triangle T$ implies that $T(H^p(X)) \subset H^q(X)$; and since $(H^p(X))^\perp = \triangle(\Omega^p(X))$, it implies also that $T((H^p(X))^\perp) \subset (H^q(X))^\perp$. It follows that

$$T \circ \pi = \pi \circ T, \tag{1}$$

and on $(H^p(X))^\perp$,

$$T \circ (\triangle | (H^p(X))^\perp) = (\triangle | (H^q(X))^\perp) \circ T, \tag{2}$$

and hence on $(H^p(X))^\perp$,

$$T \circ (\triangle | (H^p(X))^\perp)^{-1} = (\triangle | (H^q(X))^\perp)^{-1} \circ T. \tag{3}$$

It follows from (1) and (3) that $G$ commutes with $T$. □

**Theorem 9.2.6**

Each de Rham cohomology class on a compact oriented Riemannian manifold $X$ contains a unique harmonic representative.

***Proof.*** Let $\alpha$ be an arbitrary $p$-form on $X$ from the Hodge decomposition theorem and from the definition of the Green operator $G$, we have

$$\alpha = d\delta G\alpha + \delta d G\alpha + H\alpha. \tag{1}$$

Since the $G$, by Theorem 9.2.5, commutes with $d$, we have

$$\alpha = d\delta G\alpha + \delta G d\alpha + H\alpha. \tag{2}$$

Thus if $\alpha$ is a closed $p$-form

$$\alpha = d\delta G\alpha + H\alpha, \tag{3}$$

so $H\alpha$ is a harmonic $p$-form in the same de Rham cohomology class as is $\alpha$. If two harmonic forms $\alpha_1$ and $\alpha_2$ differ by an exact form $d\beta$, then we have

$$0 = d\beta + (\alpha_1 - \alpha_2).$$

But $d\beta$ and $(\alpha_1 - \alpha_2)$ are orthogonal since

$$< d\beta,\ \alpha_1 - \alpha_2 > = < \beta,\ \delta\alpha_1 - \delta\alpha_2 > = < \beta,\ 0 > = 0.$$

Thus $d\beta = 0$ and $\alpha_1 = \alpha_2$. Thus there is a unique harmonic form in each de Rham cohomology class. □

**Theorem 9.2.7**

The de Rham cohomology groups for a compact, orientable and differentiable manifold are all finite dimensional.

***Proof.*** Any differentiable manifold can be equipped with a Riemannian metric Theorem 9.2.7 follows immediately from Theorem 9.2.6 and from the finite dimensionality of the space $H^p$ of harmonic forms. □

Let $X$ be a compact, orientable and differentiable manifold of dimension $n$. We define a bilinear function

$$H^p_{deR}(X) \times H^{n-p}_{deR}(X) \to \mathbb{R} \tag{1}$$

by sending

$$(\{\omega\},\ \{\eta\}) \mapsto \int_X \omega \wedge \eta, \tag{2}$$

where $\omega$ and $\eta$ are closed forms representing the cohomology classes $\{\omega\}$ in $H^p_{deR}(X)$ and $\{\eta\}$ in $H^{n-p}_{deR}(X)$. The bilinear map (2) is well-defined. For example, if $\omega_1$ is another representative of the de Rham class $\{\omega\}$, then $\omega_1 = \omega + d\xi$ for some form $\xi$, and by the Stokes Theorem,

$$\begin{aligned}\int_X \omega_1 \wedge \eta &= \int_X \omega \wedge \eta + \int_X d\xi \wedge \eta \\ &= \int_X \omega \wedge \eta + \int_X d(\xi \wedge \eta) = \int_X \omega \wedge \eta. \end{aligned} \tag{3}$$

Also from its definition we can observe that the bilinear function (2) depends on the orientation on $X$.

**Theorem 9.2.8**

**Poincaré duality for the de Rham cohomology of a compact oriented $n$-dimensional mainfold**

The bilinear function (2) is a non-singular pairing and consequently determines isomorphisms of $H^{n-p}_{deR}(X)$ with the dual space of $H^p_{deR}(X)$, i.e.,

$$H^{n-p}_{deR}(X) \cong (H^p_{deR}(X))^*, \quad 0 \le p \le n. \tag{1}$$

***Proof.*** For given a non-zero cohomology class $\{\omega\} \in H^p_{deR}(X)$, we must find a non-zero cohomology class $\{\eta\} \in H^{n-p}_{deR}(X)$ such that $(\{\omega\},\ \{\eta\}) \neq 0$. Choose a Riemannian structure on $X$. We can assume, according to Theorem 9.2.6, that $\omega$ is the harmonic representative of $\{\omega\}$. Since the cohomology class $\{\omega\}$ is not zero, $\omega$ is not identically zero. Since $* \Delta = \Delta *$, it follows that $* \omega$ is also harmonic, and therefore closed by Theorem 9.1.3, and so $* \omega$

represents a cohomology class $\{*\omega\} \in H^{n-p}_{deR}(X)$. Now

$$(\{\omega\}, \{*\omega\}) = \int_M \omega \wedge *\omega = \|\omega\|^2 \neq 0. \tag{2}$$

Thus the pairing is non-singular, and (1) is an isomorphism. □

**Theorem 9.2.9**

If $X$ is a compact, connected, orientable, differentiable manifold of dimension $n$, then $H^n_{deR}(X) \cong \mathbb{R}$.

## 9.3 Kähler Manifold

Let $V$ be a complex vector space of complex dimension $n$. Let $V'$ be the real dual space to the underlying real vector space of $V$, and let

$$W = V' \otimes_{\mathbb{R}} \mathbb{C}$$

be the complex vector space of complex-valued real-linear mappings of $V$ to $\mathbb{C}$. Then $W$ has complex dimension $2n$, and we let

$$\wedge W = \sum_{p=0}^{2n} \wedge^p W$$

be the $\mathbb{C}$-linear exterior algebra of $W$. We will refer to an $\omega \in \wedge^p W$ as a $p$-form on $V$. Now, as before, $\wedge W$ is equipped with a natural conjugation obtained by setting, if $\omega \in \wedge^p W$,

$$\overline{\omega}(v_1, \cdots, v_p) = \overline{\omega(v_1, \cdots, v_p)}, \quad v_j \in V.$$

We say that $\omega \in \wedge^p W$ is **real** if $\overline{\omega} = \omega$, and we will let $\wedge^p_{\mathbb{R}} W$ denote the real elements of $\wedge^p W$ (noting that $\wedge^p V' \cong \wedge^p_{\mathbb{R}} W$).

Let $\wedge^{1,0}W$ be the subspace of $\wedge^{1}W$ consisting of complex-linear 1-forms on $V$, and let $\wedge^{0,1}W$ be the subspace of conjugate-linear 1-forms on $V$. Then we see that $\overline{\wedge^{1,0}W}=\wedge^{0,1}W$ and moreover

$$\wedge^{1}W=\wedge^{1,0}W\oplus\wedge^{0,1}W,$$

and this induces a bigrading on $\wedge W$,

$$\wedge W=\sum_{r=0}^{2n}\sum_{p+q=r}\wedge^{p,q}W,$$

and we see that if $\omega\in\wedge^{p,q}W$, then $\overline{\omega}\in\wedge^{p,q}W$.

We recall that a mapping $f: V\times V\rightarrow\mathbb{C}$ is **sesquilinear** if $f$ is real bilinear, and moreover, $f(\lambda u, v)=\lambda f(u, v)$, and $f(u, \lambda v)=\overline{\lambda}f(u, v)$, $\lambda\in\mathbb{C}$.

We suppose that our complex vector space is equipped with a **Hermitian inner product** $< , >$. This inner product is a Hermitian symmetric sesquilinear positive definite form, and can be represented in the following manner. If $\{z_1, \cdots, z_n\}$ is a basis for $\wedge^{1,0}W$ then $\{\overline{z}_1, \cdots, \overline{z}_n\}$ is a basis for $\wedge^{0,1}W$ and we can write, for $u, v\in V$,

$$< u, v > = h(u, v),$$

where

$$h=\sum_{\mu, \nu}h_{\mu\nu}z_{\mu}\otimes\overline{z}_{\nu},$$

and $(h_{\alpha\beta})$ is a positive definite Hermitian symmetric matrix. Now $h$ is a complex-valued sesquilinear form acting on $V\times V$, and we can write

$$h=S+iA,$$

where $S$ and $A$ are real bilinear forms acting on $V$. One finds that $S$ is a symmetric positive definite bilinear form, which represents the

**Euclidean inner product** induced on the underlying real vector space of $V$ by the Hermitian metric on $V$.

Moreover one can calculate easily that

$$A=\frac{1}{2i}\sum_{\mu,\nu}h_{\mu\nu}(z_\mu\otimes\bar{z}_\nu-\bar{z}_\nu\otimes z_\mu)$$
$$=-i\sum_{\mu,\nu}h_{\mu\nu}z_\mu\wedge\otimes\bar{z}_\nu.$$

Let us define

$$(*)\qquad \Omega=\frac{i}{2}\sum_{\mu,\nu}h_{\mu\nu}z_\mu\wedge\bar{z}_\nu,$$

the **fundamental 2-form** associated to the Hermitian metric $h$. One sees immediately that

$$\Omega=-\frac{1}{2}A=-\frac{1}{2}\operatorname{Im} h,$$

and thus

$$h=S-2i\Omega.$$

Moreover $\Omega\in\wedge_{\mathbb{R}}^{1,1}W$, i.e., $\Omega$ is a real 2-form of type (1, 1). We can always choose a basis $\{z_\mu\}$ of $\wedge^{1,0}W$ so that $h$ has the form

$$h=\sum_\mu z_\mu\otimes\bar{z}_\mu.$$

It then follows that, if we let

$$x_\mu=\frac{z_\mu+\bar{z}_\mu}{2},\quad y_\mu=\frac{z_\mu-\bar{z}_\mu}{2i}$$

be the real and imaginary parts of $\{z_\mu\}$, then

$$h=\sum_\mu(x_\mu\otimes x_\mu+y_\mu\otimes y_\mu)-2i\sum(x_\mu\wedge y_\mu),$$

and thus from ( * ), with respect to this basis,

$$S = \sum x_\mu \otimes x_\mu + y_\mu \otimes y_\mu,$$

$$(**) \qquad \Omega = \sum x_\mu \wedge y_\mu = \frac{i}{2} \sum z_\mu + \bar{z}_\mu.$$

From $(**)$ we have

$$\Omega^n = n!\, x_1 \wedge y_1 \wedge \cdots \wedge x_n \wedge y_n.$$

Thus the fundamental 2-form associated to a Hermitian metric is a real form of type (1, 1) whose coefficient matrix is positive definite, and moreover, $\Omega^n$ is a nonzero volume element of $V'$. Thus $\Omega^n$ determines an orientation on $V'$, and we see from $(**)$ that the $\{x_\mu, y_\mu\}$ is an orthonormal basis for $V'$ in the induced Euclidean metric of $V'$. Thus we see that there is a naturally defined Hodge $*$-operator

$$* : \wedge^p V' \to \wedge^{2n-p} V'$$

coming from the Hermitian structure of $V$. Namely, $V'$ has the dual metric to the real underlying vector space of $V$, while $V'$ is equipped with the orientation induced by the $2n$-form $\Omega^n$ coming from the Hermitian structure of $V$. We define

$$\text{vol} = \frac{1}{n!} \Omega^n,$$

which, with respect to the orthonormal basis used above, becomes

$$\text{vol} = x_1 \wedge y_1 \wedge \cdots \wedge x_n \wedge y_n.$$

Let $X$ be a Hermitian complex manifold with Hermitian metric $h$. Then there is, associated to $X$ and $h$, a fundamental 2-form $\Omega$, which at each point $x \in X$ is the form of type $(1, 1)$, which is the fundamental 2-form associated as in $(*)$ with the Hermitian bilinear form

$$h_x : T_x(X) \times T_x(X) \to \mathbb{C},$$

given by the Hermitian metric.

**Definition 9.3.1**

A Hermitian metric $g$ on $X$ is called a **Kähler metric** if the fundamental 2-form $\Omega$ associated with $g$ is closed; i.e., $d\Omega = 0$.

**Definition 9.3.2**

(a) A complex manifold $X$ is said to be of **Kähler type** if it admits at least one Kähler metric.
(b) A complex manifold equipped with a Kähler metric is called a **Kähler manifold**.

We shall see later that not every complex manifold $X$ admits a Kähler metric. On a complex manifold, a Hermitian metric can be expressed in local coordinates by a Hermitian symmetric tensor

$$h = \Sigma h_{\mu\nu}(z) dz_\mu \otimes d\overline{z}_\nu,$$

where $h = [h_{\mu\nu}]$ is a **positive definite Hermitian symmetric matrix** (depending on $z$); i.e., $h = {}^t\overline{h}$ and ${}^t\overline{u}hu > 0$ for all vectors $u \in \mathbb{C}^n$. The associated fundamental 2-form is then, in this notation,

$$\Omega = \frac{i}{2} \Sigma h_{\mu\nu}(z) dz_\mu \wedge d\overline{z}_\nu.$$

In the notation, the Hermitian metric is

$$h_{\mu\nu}(z) = h\left(\frac{\partial}{\partial z_\mu}, \frac{\partial}{\partial z_\nu}\right)(z).$$

Let us first give some examples of Kähler manifolds.

**Example 9.3.3**

Let $X=\mathbb{C}^n$ and let $h=\Sigma_{\mu=1}^n dz_\mu \otimes d\overline{z}_\mu$. Then

$$\Omega=\frac{i}{2}\sum_{\mu=1}^n dz_\mu \wedge d\overline{z}_\mu=\sum_{\mu=1}^n dx_\mu \wedge dy_\mu,$$

where $z_\mu=x_\mu+iy_\mu$, $\mu=1, \cdots, n$, is the usual notation for real and imaginary coordinates. Then, clearly, $d\Omega=0$, since $\Omega$ has constant coefficients, and hence $h$ is a Kähler metric on $\mathbb{C}^n$.

**Example 9.3.4**

Let $\omega_1, \cdots, \omega_{2n}$ be $2n$ vectors in $\mathbb{C}^n$ which are linear independent over $\mathbb{R}$ and let $\Gamma$ be the lattice consisting of all integral linear combinations of $\{\omega_1, \cdots, \omega_{2n}\}$. The lattice $\Gamma$ acts in a natural way on $\mathbb{C}^n$ by translation, $z \to \gamma+z$, if $\gamma \in \Gamma$. Let

$$X=\mathbb{C}^n/\Gamma$$

be the set of equivalence classes with respect to $\Gamma$, where we say that $z$ and $w$ are equivalent with respect to $\Gamma$ if $z=w+\gamma$ for some $\gamma \in \Gamma$. By giving $X$ the usual quotient topology, we see that $X$ is in a natural manner a complex manifold and that its universal covering space is $\mathbb{C}^n$. We call $X$ a **complex torus**, and $X$ is homeomorphic to $S^1 \times \cdots \times S^1$, with $2n$-factors. The Kähler metric $h$ on $\mathbb{C}^n$, given above, is invariant under the action of $\Gamma$ on $\mathbb{C}^n$; i.e., if $\gamma \in \Gamma$ gives a mapping $\gamma : \mathbb{C}^n \to \mathbb{C}^n$, then $\gamma^* h=h$, where $\gamma^*$ is the induced mapping on (covariant) tensors. Because of this invariance, we can find a Hermitian metric $\overline{h}$ on $X$ so that if

$$\pi : \mathbb{C}^n \to \mathbb{C}^n/\Gamma$$

is the holomorpic projection mapping, then $\pi^*(\overline{h})=h$. This is easy to see, and we omit any details here. Moreover, $\pi$ is a local diffeomorphism, and hence in a neighbourhood $U$ of a point $z \in \mathbb{C}^n$, we have $\pi_U := \pi|_U$ is a biholomorphic mapping. Hence $(\pi_U^{-1})^* h|_U = \tilde{h}|_{\pi(U)}$, and similarly for the corresponding $\Omega$ we have $\tilde{\Omega}$. Since $d$ commutes with $(\pi_U^{-1})^*$, we have

$$d\tilde{\Omega}|_{\pi(U)} = (\pi_U^{-1})^* d\Omega|_U = 0.$$

Then $\tilde{h}$ defined on $X$ is a Kähler metric, and all complex tori are of Kähler type.

**Example 9.3.5** One of the most important manifolds of Kähler type is $\mathbb{P}^n$. Let $(\zeta_0, \cdots, \zeta_n)$ be homogeneous coordinates for $\mathbb{P}^n$, and consider the differential form $\tilde{\Omega}$,

$$\tilde{\Omega} = \frac{i}{2} \frac{|\xi|^2 \sum_{\mu=0}^{n} d\xi_\mu \wedge d\overline{\xi_\mu} - \sum_{\mu,\nu=0}^{n} \overline{\xi_\mu} \xi_\nu d\xi_\mu \wedge d\overline{\xi_\nu}}{|\xi|^4}$$

where $|\xi|^2 = \xi_0^2 + \cdots + \xi_n^2$, as usual. This form is the homogeneous representation for the curvature form of the universal bundle over $\mathbb{P}^n$ with the standard metric on the frame bundle. In particular, then, $\tilde{\Omega}$ defines a $d$-closed differential form $\Omega$ on $\mathbb{CP}^n$ of type(1, 1). In terms of local coordinates in a particular coordinate system, for example,

$$w_j = \frac{\xi_j}{\xi_0}, \quad j = 1, \cdots, n,$$

we can write $\Omega$ as

$$\Omega(w) = \frac{i}{2} \frac{(1 + |w|^2) \sum_{\mu=1}^{n} dw_\mu \wedge d\overline{w}_\mu - \sum_{\mu,\nu=1}^{n} \overline{w}_\mu w_\nu dw_\mu \wedge d\overline{w}_\nu}{(1 + |w|^2)^2}$$

Thus the associated tensor

$$h = (\sum h_{\mu\nu}(w) dw_\mu \otimes d\overline{w}_\nu)(1+|w|^2)^{-2}$$

has coefficients (ignoring the positive denominator above)

$$h_{\mu\nu}(w) = (1+|w|^2)\delta_{\mu\nu} - \overline{w}_\mu w_\nu, \quad \mu,\ \nu = 1, \cdots, n.$$

It is easy to see that $\tilde{h} = [h_{\mu\nu}]$ is Hermitian symmetric and positive definite.

In fact, suppose that $u \in \mathbb{C}^n$. Then

$$\begin{aligned} {}^t\overline{u}\tilde{h}u &= \sum_{\mu,\nu} h_{\mu\nu} u_\mu \overline{u}_\nu = \sum_{\mu,\nu}(1+|w|^2)\delta_{\mu\nu} u_\mu \overline{u_\nu} - (\sum_\mu \overline{w_\mu} u_\mu)(\sum_\nu w_\nu \overline{u_\nu}) \\ &= |u|^2 + |u|^2|w|^2 - (\overline{w}, \overline{u})(w, u), \end{aligned}$$

letting ( , ) denote the standard inner product in $\mathbb{C}^n$. Hence by Schwarz's inequality we have

$${}^t\overline{u}\tilde{h}u \geq |u|^2,$$

and hence $\tilde{h}$ is positive definite. It then follows that $h$ defines a Hermitian metric on $\mathbb{CP}^n$ (which is called the **Fubini-Study metric** classically). Since $\Omega$ is a closed (1, 1)-form on $\mathbb{CP}^n$, as noted earlier, we see that $h$ is, in fact, a Kähler metric. This Kähler metric is invariant with respect to transformations of $\mathbb{CP}^n$ induced by unitary transformations of $\mathbb{C}^{n+1} - \{0\}$ onto itself, a property which will not concern us too much but which is important from the point of view of homogeneous spaces.

The next proposition combined with the above basic examples gives many additional examples of Kähler manifolds.

**Theorem 9.3.6**

Let $X$ be a Kähler manifold with Kähler metric $h$ and let $Y$ be a complex submanifold of $X$. Then $h$ induces a Kähler metric on $Y$, and this $Y$ becomes, therefore, a Kähler manifold.

***Proof.*** Let $j : Y \to X$ be the injection mapping. Then $h_Y = j^*h$ defines a metric on $Y$, and $j^*\Omega = \Omega_Y$ is the associated fundamental form to $h_Y$ on $Y$. Since $d\Omega_Y = j^*d\Omega = 0$, it is clear that $\Omega_Y$ is also a Kähler fundamental 2-form. □

## 9.4 Hodge Decomposition Theorem on Compact Kähler Manifold

In terms of the differential operators $d$, $\partial$, and $\bar{\partial}$ on a Hermitian manifold, we can define the following Laplacian operators,

$$\triangle = dd^* + d^*d$$
$$\square = \partial\partial^* + \partial^*\partial$$
$$\overline{\square} = \bar{\partial}\bar{\partial}^* + \bar{\partial}^*\bar{\partial},$$

of which the first and last play an important role in our study of the Kähler manifold. Note that $\overline{\square}$ is the complex conjugate of the operator $\square$, thus justifying the notation. What can we say about the relation between the Laplacians $\triangle$ and $\overline{\square}$? In general not too much, but on Kähler manifolds there is a relationship (Theorem 9.4.2). We assume that Theorem 9.4.1 is true.

**Theorem 9.4.1**

Let $X$ be a Kähler manifold. If the differential operators $d$, $d^*$, $\bar{\partial}$, $\bar{\partial}^*$, $\partial$, $\partial^*$, $\square$, $\overline{\square}$, and $\triangle$ are defined with respect to the Kähler metric on $X$, then $\triangle$ commutes with $*$, $d$, and $L$, and

$$\Delta = 2\square = 2\overline{\square}.$$

Let $X$ be a compact complex manifold. Then we have the de Rham cohomology groups on $X$, $\{H^*(X, \mathbb{C})\}$, represented by $d$-closed differential forms with complex coefficients, and the Dolbeault cohomology groups on $X$ $\{H^{p,q}(X)\}$, represented by $\bar{\partial}$-closed $(p, q)$-forms. We have seen that these vector spaces are finite dimensional. Moreover, there is a spectral sequence relating them. However, in general, if $\varphi$ is a $\bar{\partial}$-closed $(p, q)$-form on $X$, then $\varphi$ need not be $d$-closed, and, conversely, if $\Psi$ is a $d$-closed $r$-form on $X$ and

$$\Psi = \Psi^{r,0} + \Psi^{r-1,1} + \cdots + \Psi^{0,r}$$

are the bihomogeneous components of $\Psi$, then the components $\Psi^{p,q}$ need not be $\bar{\partial}$-closed. On manifolds of Kähler type, however, such relations are valid, as we see in the following decomposition theorem of Hodge.

**Theorem 9.4.2 Hodge**

Let $X$ be a compact complex manifold of Kähler type. Then there is a direct sum decomposition

$$(1) \qquad H^r(X, \mathbb{C}) = \sum_{p+q=r} H^{p,q}(X),$$

and moreover

$$(2) \qquad \overline{H^{p,q}}(X) = H^{q,p}(X).$$

***Proof.*** We shall show that

$$H^r(X) = \sum_{p+q=r} H^{p,q}(X),$$

and then (1) follows immediately. Suppose that $\varphi \in H^r(X)$. Then $\Delta\varphi = 0$, but $2\overline{\Box} = \Delta$, by Theorem 9.4.1, and hence $\overline{\Box}\varphi = 0$. But $\varphi = \varphi^{r,0} + \cdots + \varphi^{0,r}$ (writing out the bihomogenous terms), and, moreover,

$$\overline{\Box}\varphi = \overline{\Box}\varphi^{r,0} + \cdots + \overline{\Box}\varphi^{0,r}.$$

Since $\overline{\Box}$ preserves bidegree, we see that $\overline{\Box}\varphi = 0$ implies that $\overline{\Box}\varphi^{r,0} = \cdots \overline{\Box}\varphi^{0,r} = 0$, and therefore there is a mapping

$$\tau : H^r(X) \to \sum_{p+q=r} H^{p,q}(X)$$

given by $\tau(\varphi) := (\varphi^{r,0}, \cdots, \varphi^{0,r})$. The mapping is clearly injective, and moreover, if $\varphi \in H^{p,q}(X)$, then $\overline{\Box}\varphi = \frac{1}{2}\Delta\varphi = 0$, which implies that $\varphi \in H^{p+q}(X)$, and thus $\tau$ is surjective.

Assertion (2) follows immediately from the fact that $\overline{\Box}$ is real by Theorem 9.4.1 and that conjugation is an isomorphism from $\Omega^{p,q}(X)$ to $\Omega^{q,p}(X)$. $\Box$

As a consequence of the Hodge decomposition theorem, we have the following relations for the Betti numbers and Hodge numbers of a Kähler manifold. Recall that

$$b_r(X) = \dim_{\mathbb{C}} H^r(X, \mathbb{C}),$$
$$h^{p,q}(X) = \dim_{\mathbb{C}} H^{p,q}(X).$$

**Theorem 9.4.3**

Let $X$ be a compact Kähler manifold. Then we get

(1) $b_r(X) = \Sigma_{p+q+r} h^{p,q}(X)$.

(2) $h^{p,q}(X) = h^{q,p}(X)$.

(3) $b_q(X)$ is even for $q$ odd.

(4) $h^{1,0}(X) = \frac{1}{2} b_1(X)$ is a topological invariant. □

These results are simple consequences of Theorem 9.4.2.

The simplest example of a non-Kähler compact complex manifold is given by a **Hopf surface** $X$. The surface $X$ is a compact complex manifold which is homeomorphic to $S^1 \times S^3$. Thus the Betti numbers $b_1(X) = b_3(X) = 1$ and so is non-Kähler.

# Appendix

# Appendix A: Characteristic Class

The theory of characteristic classes began in 1935 Stiefel-Whitney classes by Hassler Whitney (Institute for Advanced study in Princeton, sphere bundle) and Eduard Stiefel (director of the Institute for Applied Mathematics of the Federal Institute of Technology in Zürich, tangent bundle), in 1942 Pontrjagin classes by Lev Pontrjagin (director of the Steklov Institute in Moscow, Grassmann manifold), and in 1949 Chern classes by Shing-Shen Chern (the University of California in Berkeley, complex Grassmann manifold). In this Appendix A we would like to survey charecteristic classes.

## §1. Vector Bundle

Let $F$ be a field, and let $\pi : E \rightarrow X$ be a **vector bundle** over $X$ with total space $E$, fibre over $x \in X$ a $F$-vector space $E_x$ of dimension $n$, the set of zeros $Z$, and projection map $\pi$. Consider the equivalence relation $\rho$ defined on $E-Z$ by : $e_1$, $e_2 \in E$ are equivalent iff $\pi(e_1) = \pi(e_2) = x \in X$ and, as vectors in $E_x$, $e_1 = \lambda e_2$ for some $\lambda \in F$.

The space $\widetilde{E} = (E-Z)/\rho$ of equivalence classes, with the map

$$\widetilde{\pi} : \widetilde{E} \rightarrow X$$

induced by $\pi$, is called the **projective bundle associated to** $E$. The projection map $p : (E-Z) \rightarrow \widetilde{E}$ defines a line bundle $L_E$ over $\widetilde{E}$. $L_E$ is a subbundle of $\widetilde{\pi}^* E$ with a quotient bundle $Q$ of rank $n-1$. Sometimes $L_E$ is called the '**canonical' subbundle** of $\widetilde{\pi}^* E$; on each fibre of $\widetilde{E}$ it is the well known **Hopf bundle**. A map $f : Y \rightarrow X$ defines an induced bundle $E_1 = f^* E$ and a map $\widetilde{f} : \widetilde{E_1} \rightarrow \widetilde{E}$ such that $L_{E_1} = \widetilde{f}^* L_E$.

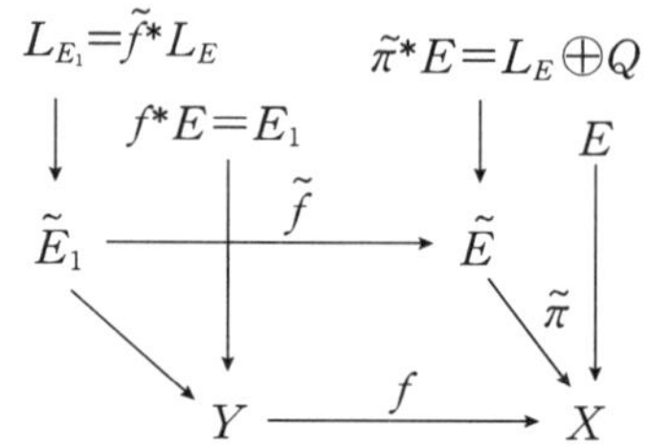


Let $\mathscr{C}$ be the category of topological spaces. For each topological space $X\in\mathscr{C}$, let $\mathcal{E}(X)$ be the vector bundles over $X$,

$$A(X) = \underset{i=0}{\oplus} A^i(X)$$

a graded ring with unit such that

$$A^i(X)A^j(X) \subset A^{i+j}(X),$$

for $f: Y\to X$, $A(f)=f^*: A(X)\to A(Y)$.

Then the $A$ satisfies the following (A1)~(A4):

**(A1)** If $f: Y\to X$ is in $\mathscr{C}$ and $E\in\mathcal{E}(X)$ then $E_1=f^*E\in\mathcal{E}(Y)$.

**(A2)** If $E\in\mathcal{E}(X)$, $X\in\mathscr{C}$, then $\tilde{\pi}: \tilde{E}\to X$ is in $\mathscr{C}$ and $L\in\mathcal{E}(\tilde{E})$.

**(A3)** In the above notations, the map $\tilde{f}: \tilde{F}\to\tilde{E}$ is in $\mathscr{C}$.

**(A4)** A is a contravariant functor i.e., maps $Z\xrightarrow{g}Y\xrightarrow{f}X$ in $\mathscr{C}$ define homomorphisms

$$A(X)\xrightarrow{f^*}A(Y)\xrightarrow{g^*}A(Z)$$

which preserve grading and unit and for which $(fg)^*=g^*f^*$.

## §2. Characteristic Class for Line Bundle

Let $\mathscr{L}(X)$ be the subset of $\mathcal{E}(X)$ consisting of the bundles of rank 1 (if the fibre at $x$, $E_x = F^p$ we write the **rank** $\mathrm{rk}\,(E) = p$). A bundle in $\mathscr{L}(X)$ is called a **line bundle**. The $\mathscr{L}(X)$ is a group with multiplication $\otimes$ defined by the tensor product. The unit in $\mathscr{L}(X)$ is the trivial bundle $I = X \times F$. The inverse of $L$ is the dual bundle $L^*$.

We assume that it is possible to attach a meaning to the phrase "a section $s$ of $E$ is transversal to the zero section". In the case of differential analytic and non-singular algebraic varieties and transversality may be defined in terms of tangent spaces. In other cases it requires a suitable intersection theory. We assume as part of the definition that the set of zeros of such a section is in $\mathscr{C}$. We say that **characteristic classes** $\gamma(L)$ for line bundles $L$ are defined by

**(B1)** There is a group homomorphism

$$\gamma : \mathscr{L}(X) \to A'(X), \ \gamma(L_1 \otimes L_2) = \gamma(L_1) + \gamma(L_2),$$

such that if $f : Y \to X$ is in $\mathscr{C}$ and $L \in \mathscr{L}(X)$ then

$$f^* \gamma(L) = \gamma(f^* L).$$

**(B2)** If $s$ is a section of $L$ which is transversal to the zero section, and $j : Y \to X$ is the inclusion map of the set of zeros of $L$, then $j^* a = 0$, $a \in A(X)$ implies $a \cdot \gamma(L) = 0$.

(B1) and (B2) are consistent with (A1)~(A4) because they are satisfied if $\gamma(L) = 0$ for all line bundles $L$. This case does occur in connection with the Euler class: see section §5 of Appendix A, Theorem (d).

In §3 of Appendix A we show that line bundle, one additional axiom and characteristic classes can be defined for vector bundles.

## §3. Chern Class

Suppose that the ring $A(X)$ is commutative and that, in addition to A(1)~A(4) and B(1)~B(2) the functor $A$ and the classes $\gamma(L)$ satisfy

**(C1)** Let $E \in \mathcal{E}(X)$ and $\xi_E = \gamma(L_E^*)$, $\tilde{\pi}^* E = L_E \oplus Q$. Then $A(\tilde{E})$ is a left module over $A(X)$ via $\tilde{\pi}^* : A(X) \to A(\tilde{E})$ and the elements $\xi_E^i$, $0 \le i \le p-1$, $\mathrm{rk}(E) = p$, form a base for $A(\tilde{E})$ over $A(X)$.

(C1) implies that $\tilde{\pi}^*$ is a monomorphism.

(C1) implies that there is a relation of the form

$$\xi_E^p + \xi_E^{p-1} \tilde{\pi}^* c_1(E) + \cdots + \tilde{\pi}^* c_p(E) = 0 \tag{1}$$

where $p = \mathrm{rk}(E)$ and $c_i(E) \in A^i(X)$ are well defined because $\tilde{\pi}^*$ is a monomorphism.

**Definition 1**

$c(E) = 1 + \sum_{i=1} c_i(E) \in A(X)$ is the **total Chern class** of $E$.

**Lemma 1**

If $E$ has a never zero section, $\mathrm{rk}(E) = p$, then $c_p(E) = 0$.

***Proof.*** The section defines a map $s : X \to \tilde{E}$ and $s^* L_E = I$ the trivial bundle. By (B1),

$$s^* \xi_E = s^*(\gamma(L_E)) = \gamma(s^*(L_E)) = \gamma(I) = 0.$$

Applying $s*$ to (1) we obtain

$$s^* \tilde{\pi}^* c_p(E) = c_p(E) = 0. \qquad \square$$

**Lemma 2**

If $E_1 = E \otimes L^*$, $L \in \mathscr{L}(X)$, $\gamma(L^*) = \lambda$, $\mathrm{rk}(E) = p$,

then $c_p(E_1) = \sum_{i=0}^{p} \lambda^{p-i} c_i(E)$.

***Proof.*** The definitions of $\widetilde{E}$, $L_E$ imply that $\widetilde{E}_1 = \widetilde{E}$ but that $L_{E_1} = L_E \otimes \widetilde{\pi}^* \check{L}$, and so by (B1)

$$\xi_E = \xi_{E_1} + \widetilde{\pi}^* \lambda.$$

Substituting this value in (1) gives the required result. □

**Lemma 3**

Suppose there is an exact sequence of vector bundles on $X$

$$0 \to E_1 \to E \to L \to 0$$

with $\mathrm{rk}(E) = p$, $\mathrm{rk}(L) = 1$ and $\gamma(L) = \lambda$. Then $c(E) = c(E_1) \cdot (1 + \lambda)$

***Proof.*** Regard $\widetilde{E}_1$ as a subspace of $\widetilde{E}$, with an injection map $j$. Then by (B1), $j^* \xi_E = \xi_{E_1}$. The epimorphism $E \to L$ defines an epimorphism $\widetilde{\pi}^* E \to \widetilde{\pi}^* L$ and therefore a homomorphism $L_E \to \widetilde{\pi}^* L$ the set of zeros of which is $\widetilde{E_1}$. The section of $L_E^* \otimes \widetilde{\pi}^* L$ so defined is transverse to the zero section, and has the set $(T)$ of zeros $\widetilde{E}_1$. Since

$$\xi_{E_1}^{p-1} + \xi_{E_1}^{p-2} j^* \widetilde{\pi}^* c_1(E_1) + \cdots + j^* \widetilde{\pi}^* c_{p-1}(E_1) = 0,$$

we have

$$j^* (\xi_E^{p-1} + \xi_E^{p-2} \widetilde{\pi}^* c_1(E_1) + \cdots + \widetilde{\pi}^* c_{p-1}(E_1)) = 0$$

and, by (B1) and (B2),

$$(\xi_E+\tilde{\pi}^*\lambda)(\xi_E^{p-1}+\xi_E^{p-2}\tilde{\pi}^*c_1(E_1)+\cdots+\tilde{\pi}c_{p-1}(E_1))=0$$

and so $c(E)=(1+\lambda)c(E_1)$.

Here and in the sequel we have marked $(T)$ those points where a statement of transversality must be checked in the category concerned. □

**Theorem 1**

The total Chern class of $E\in\mathcal{E}(X)$

$c(E)=1+\sum_{i=1}^{p}c_i(E)$ satisfies

(a) $f^*c(E)=c(f^*E)$ for $f: Y\to X$ in $\mathscr{C}$, $E$ in $\mathcal{E}(X)$

(b) $c(L)=1+\lambda(L)$ for $I\in\mathscr{L}(X)$

(c) An exact sequence $0\to E_1\to E\to Q\to 0$

implies $c(E)=c(E_1)\cdot c(Q)$.

***Proof.*** (a) Since $\xi_{E_1}=f^*\xi_E$, (B1) implies that

$$\sum_{i=0}^{p}\tilde{\pi}^*(f^*c_i(E))\cdot\xi_{E_1}^{p-i}=\sum_{i=0}^{p}f^*\tilde{\pi}^*c_i(E)\cdot f^*\xi_E^{p-i}$$
$$=f^*\left(\sum_{i=0}^{p}\tilde{\pi}^*c_i(E)\cdot\xi^{p-i}\right)=0,$$

and so $f^*c(E)=c(f^*E)$.

(b) If $p=1$ we have $\tilde{L}=X$, $L_L=L$, $\tilde{\pi}=id$ and therefore

$$c_1(L)=-\xi_L=-\gamma(L^*)=\gamma(L).$$

(c) Recall that a vector bundle $E$ is completely reducible iff there is a sequence $\{E_i\}$, $1\le i\le p$, of subbundles of $E$, with

$$rk(E_i)=i,\ E=E_p,\ L_i=E_i/E_{i-1}.$$

Lemma 3 implies that, if $E$ is completely reducible, then

$$c(E) = c(L_1) \cdots c(L_p).$$

If $F$ and $Q$ are completely reducible, then so is $E$, so that (c) is proved in this case. To complete the proof it is sufficient to prove Lemma 4. □

**Lemma 4**

If $E \in \mathcal{E}(X)$, then there is a map $f: Y \to X$ such that $f^*E$ is completely reducible and

$$f^*: A(X) \to A(Y)$$

is a monomorphism.

***Proof.*** We suppose that there is a sequence $\{E_i\}$, $1 \le i \le r$, of subbundles of $E$, with $\mathrm{rk}(E_i) = i$. Then there is an exact sequence

$$0 \to E_i \to E \to Q \to 0$$

with $\mathrm{rk}(Q) = p - i$. Let $\tilde{\pi}: \widetilde{Q} \to X$ be the projective bundle associated to $Q$. There is an exact sequence

$$0 \to \tilde{\pi}^* E_i \to \tilde{\pi}^* E \to \tilde{\pi}^* \widetilde{Q} \to 0$$

and by (C1), $\tilde{\pi}^*$ is a monomorphism. Let $E_{i+1}$ be the subbundle of $\tilde{\pi}^* E$ consisting of those vectors which map onto $L_Q \subset \tilde{\pi}^* Q$. The sequence $\{\tilde{\pi}^* E_r, E_{i+1}\}$ of subbundles of $\tilde{\pi}^* E$ has $\mathrm{rk}(\tilde{\pi}^* E_r) = r$, $\mathrm{rk}(E_{i+1}) = i + 1$. Repeating this construction $p$ times gives a map $f$ with the required property. □

**Lemma 5**

For the dual $E^*$ of $E$ we have

$$c_i(E) = (-1)^i c_i(E^*).$$

***Proof.*** Use Lemma 4 to reduce $L_i$ to the case where $E$ and $\breve{E}$ are both completely reducible. Then if $c(E) = \Pi(1+\gamma(L_i))$,

$$c(E^*) = \Pi(1-\gamma(L_i))$$

which gives the result. □

Another application of Lemma 4 is to the uniqueness of Chern classes, in fact, given a theory of characteristic classes for line bundles, the Chern classes are determined uniquely by (a), (b), (c) of Theorem 1.

**Theorem 2**

Suppose $d(E) = 1 + \sum_{i=1} d_i(E)$, $d_i(E) \in A^i(X)$, satisfies (a), (b), (c) of Theorem 1. Then $d(E) = c(E)$.

***Proof.*** Proceed by induction on $\mathrm{rk}(E) = p$.

When $p = 1$, (b) implies that $d(E) = c(E)$.

Suppose that $d(Q) = c(Q)$ for all $Q$ of rank $p-1$ and consider the exact sequence

$$0 \to L_E \to \tilde{\pi}^* E \to Q \to 0$$

of vector bundles on $\tilde{E}$. Then (a) and (c) imply

$$\tilde{\pi}^* d(E) = d(Q) \cdot d(L_E) = c(Q) \cdot c(L_E) = \tilde{\pi}^* c(E)$$

and (C1) implies that $\tilde{\pi}^*$ is a monomorphism.

Thus we have $d(E) = c(E)$. □

Notice the following properties of the top Chern class. If $\mathrm{rk}(E) = p$ there is a **Euler class** $e(E) = c_p(E) \in A^p(X)$ with the properties

(a) $f^* e(E) = e(f^* E)$ for $f: Y \to X$ in $\mathscr{C}$, $E$ in $\mathcal{E}(X)$,

(b) $e(L) = \gamma(L)$ for $L \in \mathscr{L}(X)$,

(c) An exact sequence $0 \to F \to E \to Q \to 0$ implies

$$e(E) = e(F) \cdot e(Q).$$

The same argument as above implies that if $e(E) \in A^p(X)$ is any other class satisfying (a), (b), (c) then, using (C1), we must have $e(E) = c_p(E)$.

## §4. Gysin Homomorphism

In geometrical applications the characteristic classes are often regarded as subspaces of the space $X$ by Poincaré duality. That is, they are homology, instead of cohomology, classes. We state the properties of Poincaré duality to be used in terms of a Gysin homomorphism.

By "an inclusion map $j: Y \to X$" we mean an inclusion map in $\mathscr{C}$ with $Y$ closed in $X$.

Let $E \in \mathcal{E}(X)$ and consider the projective bundle associated to $E \otimes I$. We denote this bundle by $E'$, its projection map by $\pi' : E' \to X$, and the canonical line subbundle of $\pi'^* E \oplus I$ by $L_{E'}$. A section $s$ of $E$ defines an inclusion map $s' : X \to E'$. Then we have $s'^* L_{E'} = I$.

**(D1)** An inclusion map $i : Y \to X$ defines a group homomorphism $i_* : A(Y) \to A(X)$ which raises degree by the codimension of $Y$ in $X$, $p = \mathrm{codim}_X Y$. If $j : Z \to Y$ is also an inclusion map then $(ij)_* = i_* j_*$.

**(D2)** If $a \in A(X)$, $b \in A(Y)$ then $i_*(i^* a, b) = a \cdot i_* b$.

**(D3)** If $L_1, L_2 \in \mathscr{L}(X)$. Let $z_1, z_2$ be zero sections of $L_1, L_2$, let $z_1, z_2$ be zero sections of $L_1, L_2$, let $s_1, s_2$ be arbitrary sections of $L_1, L_2$, and let $z = z_1 \otimes z_2$, $s = s_1 \otimes s_2$ be the corresponding sections of $L = L_1 \otimes L_2$. Then

$$s'^* z'_* 1 = s_1'^* z_1'^* 1 + s_2'^* z_2'^* 1.$$

Using (D1)~(D3) we shall, by defining transversality and $\gamma(L)$ suitably, recover the characteristic classes of line bundles of section §2 of Appendix A.

### Definition 2

A map $f: Y \to X$ is **transversal** to $M \subset X$ if

(i) the space $N = f^{-1}(M)$,

and the inclusion maps $i: M \to X$, $j: N \to Y$ are all in $\mathscr{C}$, and

(ii) $f^* i_* 1 = j_* 1$.

### Definition 3

A section $s$ of $E$ is **transversal** to $z'(X) \subset E'$, where $z$ is the zero section of $E$.

We shall use transversality only in the following simple cases, which can easily be checked in any given category.

(a) Let $F = f^* E$ be induced by a map $f: Y \to X$, and let $f': F' \to E'$ be the corresponding map of projective bundles. Then $f'$ is transversal to $z'(X) \subset E'$.

(b) If $E = F \oplus Q$ the inclusion maps $F' \subset E'$, $Q' \subset E'$ are transversal.

(c) If $E$, $F \in \mathcal{E}(X)$, $\mathrm{rk}(F) = q$, and $s$ is a section of $L_{E'}{}^* \otimes \pi'^* \times F$ which on each fibre of $E'$ has a set of zeros a linear subspace of codimension $q$, then $s$ is transversal to the zero section.

**Theorem 3**

Axioms (D1)~(D3) imply the existence of characteristic classes for line bundles.

***Proof.*** It is necessary to define $\gamma(L)$ so that B(1), B(2) hold. If $\gamma(L) =$

$z'^{*}z'_{*}1$ then (D3) with $s_1 = z_1$, $s_2 = z_2$ implies that $\gamma$ is a homomorphism

$$\gamma: \mathscr{L}(X) \to A^1(X).$$

If $f' : (f^*L)' \to L'$ is the map of projective bundle induced by a map $f : Y \to X$ and if $w$, $z$ are the zero sections of $f^*L$, $L$, then transversality implies that

$$\begin{aligned} f_\gamma^*(L) &= f_{z'}^{*}{}^* z'_* 1 \\ &= w'^* f'^* z'_* 1 \\ &= w'^* w'_* 1 \\ &= \gamma(f^* L) \end{aligned}$$

and so (B1) holds. To prove (B2), let $s$ be a section of $L$ transversal to the zero section, and suppose that $j : Y \to X$ is its set of zeros. Then $j_* 1 = s'^* z'_* 1$. Taking $L_1 = I$, $s_1 = z_1$, in (D3), gives $s'^* z'_* 1 = \gamma(L)$ and therefore $j^* 1 = \gamma(L)$. Suppose that $a \in A(X)$ is such that $j^* a = 0$. Then (D2) gives

$$\begin{aligned} a \cdot \gamma(L) &= a \cdot j_* 1 \\ &= j_*(j^* a) \\ &= j_*(0) = 0. \end{aligned}$$

We shall from now on assume that $\gamma(L)$ is defined in the above way. □

## Definition 4

Let $E \in \mathcal{E}(X)$, and let $z' : X \to E'$ be the inclusion map induced by the zero section of $E$. The Euler class of $E$ is $e(E) = z'^* z'_* 1$.

**Theorem 4**

The Euler class satisfies

(a) $e(f^*(E)) = f^*(e(E))$ for $f: Y \to X$ in $\mathscr{C}$

(b) $e(L) = \gamma(L)$ for $L \in \mathscr{L}(X)$

(c) $e(F \oplus G) = e(F) \cdot e(G)$.

***Proof.*** (a) follows from the transversality condition used already for line bundles in the proof of Theorem 3.

(b) is the definition of $\gamma(L)$. To prove (c) consider the inclusions $u: F' \to E'$, $v: G' \to E'$ defined by the direct sum $E = F \oplus Q$. If $f': X \to F'$, $g': X \to G'$, $z': X \to E'$ are the inclusions induced by the zero sections then

$$\begin{aligned} z'_*1 &= u_* f'_* 1 \\ &= u_* u^* g'_* 1 \\ &= u_* 1 \cdot v_* 1 \end{aligned}$$

and therefore $z'^* z'_* 1 = z'^* u_* 1$,

$$\begin{aligned} z'^* v_* 1 &= g'^* v^* u_* 1 \cdot f'^* u^* v_* 1 \\ &= g'^* g'_* 1 \cdot f'^* f'_* 1 . \end{aligned}$$

In Theorem 3 and 4 we have not assumed (C1). We shall do so in Theorem 5 to obtain an important geometrical property of the top Chern class which identifies it with the Euler class. If $E \in \mathcal{E}(X)$ has rank $p$, and $z': X \to E'$ is the inclusion map induced by the zero section, then $\xi = \gamma(L^*_{E'})$. Since $z'^* L_{E'} = I$, $z'^* \xi = 0$. □

**Theorem 5**

$$z'_* 1 = \sum_{i=0}^{p} \pi'^* c_i(E) \xi^{p-i}.$$

***Proof.*** Assume that the theorem is proved for bundles of rank strictly less than $p$, and suppose that there is an exact sequence

$$0 \to F \to E \to L \to 0.$$

The map $w : X \to F'$ induced by the zero section of $F$ and the inclusion $j : F' \to E'$ compose to give the map $z = jw : X \to E'$. Our assumption implies that

$$\begin{aligned} z'_* 1 &= j_* w_* 1 \\ &= j_* \left( \sum_{i=0}^{p-1} j^* \pi'^* c_i(F) \cdot j^* \xi^{p-1-i} \right) \\ &= \left( \sum_{i=0}^{p-1} \pi'^* c_i(F) \cdot \xi^{p-1-i} \right) \cdot (j_* 1). \end{aligned}$$

But $j_* 1 = \gamma(L_{E'}^* \otimes \pi^* L)$ so that, if $\gamma(L) = \lambda$,

$$\begin{aligned} z'_* 1 &= \left( \sum_{i=0}^{p-1} \pi'^* c_i(F) \xi^{p-1-i} \right) (\xi = \pi'^* \lambda) \\ &= \sum_{i=0}^{p} \pi'^* c_i(E) \xi^{p-i}. \end{aligned}$$

We have therefore proved the required formula under the hypothesis: $E$ has a quotient line bundle. By the argument of Lemma 4 there is a map $f : Y \to X$ such that $f^* E$ does have this property and such that $f^*$ is a monomorphism. The result now follows from the fact that $f'$ is transversal to $z'(X)$. □

**Corollary 1**

$e(E) = c_p(E)$

***Proof.*** Since $z'^* \xi = 0$, the result follows applying $z'^*$ to the formula of Theorem 4. □

**Corollary 2** Let $s$ be a section of $E$ transversal to the zero section, with set of zeros $j: Y \to X$. Then $e(E) = c_p(E) = j_* 1$.

***Proof.*** By the same argument as above, $s^* \xi = 0$ and so $s'^* z'_* 1 = c_p(E)$. Since $s$ is transversal to the zero section, this is also $j_* 1$.

It is interesting to note that Corollary 2 in its turn implies Theorem 5, and even a little more! In view of Lemma 2, the result of Theorem 5 may be stated

$$z'_* 1 = c_p(L_{E'}^* \otimes \pi'^* E).$$

□

Corollary 2 implies further that, for any section $s$,

$$s'_* 1 = c_p(L_{E'}^* \otimes \pi'^* E).$$

This is because $s$ defines an exact sequence of bundles on $E'$

$$0 \to I \to \pi'^* E \oplus I \to \pi'^* Q \to 0$$

with $c(Q) = c(E)$, and hence a map $L_{E'} \to \pi'^* Q$ which is zero on $s'_* 1$. The corresponding section of $L_{E'}^* \otimes \pi'^* Q$ is transversal to the zero section and

$$\begin{aligned} s'_* 1 &= c_p(L_{E'}^* \otimes \pi'^* Q) \\ &= c_p(L_{E'}^* \otimes \pi^* E). \end{aligned}$$

**Corollary 3** Let $u$, $s$ be arbitrary sections of $E$. Then $e(E) = u'^* s'_* 1$.

## §5. Euler Class

Axiom (C1) was stated on the assumption that $A(X)$ was commutative. For example, in the category of real vector bundles (C1) holds for

$$A^i(X) = H^i(X, \mathbb{Z}_2),$$

giving the Stiefel-Whitney classes. On the other hand, the ring

$$H^*(X, \mathbb{Z}) = \oplus H^i(X, \mathbb{Z})$$

is anticommutative, and so (C1) cannot hold if $A^i(X) = H^i(X, \mathbb{Z})$. For complex or quaternion vector bundles, this difficulty does not arise since $A^i(X)$ is then $H^{2i}(X, \mathbb{Z})$ or $H^{4i}(X, \mathbb{Z})$ respectively. In these cases (C1) holds, for the projective space $P$ has the cohomology generated by an element in $A^1(P)$.

Axioms (A1)~(A4) and (D1)~(D3) still hold for real vector bundles $E$ and anticommutative $A(X)$ provided that the bundle $E$ is oriented. Then $\widetilde{E}$ is orientable. In the absence of (C1) it is not possible to derive all the characteristic classes from the characteristic classes for line bundles. Although the Euler class still exists, it is not characterized uniquely by properties (a), (b), (c) of Theorem 4.

In this section is sketched a treatment of the Euler class based on quite different construction, exploiting not an associated projective bundle, but the associated sphere bundle.

Let $\mathcal{E}(X)$ be the category of real oriented vector bundles on $X$. If $E \in \mathcal{E}(X)$, $\mathrm{rk}(E) = p$, denote by the disc bundle $(E', E_o')$ the pair of spaces defined by vectors of length $\leq 1$, $= 1$ respectively. $E_o'$ is a $(p-1)$-sphere bundle over $X$, and the zero section defines an inclusion map $z : X \to E'$ which is disjoint from $E_o' \subset E'$.

Assume $H^*(X) = \oplus H^i(X)$ is the anticommutative cohomology ring, and that the relative cohomology group $H^*(E', E_o')$ is also defined.

**(E1)** The homomorphism

$$z_* : H^i(X) \to H^{i+p}(E', E_o')$$

is an isomorphism for all $i$. If $f: Y \to X$ is a map inducing

$$F = f^* E, \quad f' : (F', F_o') \to (E', E_o'),$$

then $z_* f^* 1 = f^* z_* 1$.

***Remark***

The existence of $z_*$ follows from the isomorphism

$$H^{i+p}(E', E_o') = H^{i+p}(T(E))$$

where $T(E)$ is the **Thom space** of $E$, obtained from $E'$ by identifying $E_o'$ to a point. The last statement of (E1) is then a transversality statement of the kind already used. The essential result, that $z_*$ is an isomorphism may, like (C1), be obtained from a Spectral Sequence argument.

## Definition 5

$e(E) = z^* z_* 1 \in H^p(X, \mathbb{Z})$ is the **Euler class** of $E$.

The corresponding element in rational/real cohomology is the rational/real Euler class of $E$.

**Theorem 6**

(a) $e(f^* E) = f^* e(E)$ which follows from $E_1$.

(b) $e(L) = \gamma(L) = 0$ because we are restricted to oriented bundles and so $L = I$ and $\gamma(L) = z^* z_* 1 = 0$.

(c) $e(F \oplus Q) = e(F) \cdot e(Q) = e(Q) \cdot e(F)$ follows either by proving a transversality condition as in Theorem 4, or by checking that $e(F) \cdot e(Q)$ satisfies $(f)$ below when $E = F \oplus Q$.

(d) If $p$ is odd then $2e(E) = 0$ and therefore the rational and real Euler classes are zero.

**Proof.** $z_*2e(E)=2z_*z^*z_*1=2(z_*1)(z_*1)=0$ because $H^*(X)$ is anticommutative. Axiom (E1) implies that $z_*$ is a monomorphism and therefore $2e(E)=0$. □

**Theorem 6**

(e) Let $\pi: E_0' \to X$ be the projection map. $e(E)$ generates the kernel of $\pi^*: H^p(X) \to H^p(E_0')$, $p=\mathrm{rk}(E)$.

**Proof.** Consider the exact sequence of cohomology

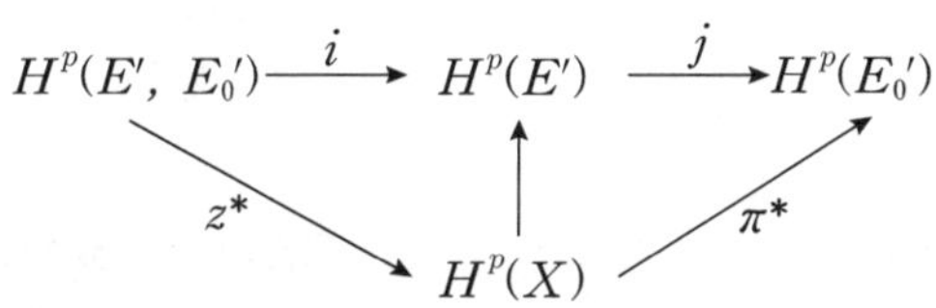


where the triangles are commutative, the map $H^p(X) \to H^p(E')$ is an isomorphism, and $\pi: E_0' \to X$ is the projection map. The exactness implies

(1) $ji\,(z_*1)=0$ and so $\pi^* e(E)=0$.

(2) If $\pi^* e(E)=0$ then $x=z^*y$, for some $y \in H^p(E', E_0')$. Axiom (E1) implies that $y=nz_*1$ for some integer $n$. □

**Theorem 6**

(f) Let $k$: $(D^p, S^{p-1}) \to (E', E_0')$ be the inclusion map of the disc and sphere over a point $x_0 \in X$. Then (E1) gives $k^*z_*1=w_*1$, where $w: x_0 \to S^p$ is the map induced by the zero section. Since the bundle is oriented, $w_*1$ is just the distinguished generator in $H^p(S^p)$. Conversely, if $e \in H^p(X)$ is of the form $z^*U$, and $k^*U = w_*1$, then $e$ must be the Euler class.

(g) Any section $s$ of $E$ defines an inclusicn map $X \to T(E)$ and therefore maps

$$s_*: H^i(X) \to H^{i+p}(E', E_0'),$$
$$s^*: H^i(E', E_0') \to H^i(X).$$

Suppose that by homotopy these homomorphisms are independent of the choice of $s$. Then, if $s$, $u$ are arbitrary sections of $E$,

$$e(E) = u^* s_* 1.$$

Suppose $s$ is transversal to the zero section, and that $j: Y \to X$ is its set of zeros. Then we deduce

$$e(E) = j_* 1.$$

Comparing with Corollary 2 of Theorem 5 this is consistent with our previous definition of the Euler class.

(h) Let $E$ be a complex vector bundle and define $e(E)$ to be the Euler class of the real vector bundle obtained by ignoring the complex structure but keeping the orientation.
Then (h) and the argument of Lemma 3 and Theorem 2 can be used to show that $e(E) = c_p(E)$, that is, the Euler class is the top Chern class.

(i) In a similar way, $e(E)$ reduced mod 2 is the top Stiefel-Whitney class. Note that if we attempted the construction of the Euler class for unoriented vector bundles, $z_*$ would be defined only in mod 2 cohomology, so that $e(E)$ would lie in $H^p(X, \mathbb{Z}_2)$, and just be the top Stiefel-Whitney class.

## §6. Pontrjagin Class

Let $E$ be a real vector bundle on $X$, and let $E^{\mathbb{C}}$ be the complexification of $E$, that is, the complex vector bundle with the coordinate transformations as $E$.

Define

$$p_i(E)=(-1)^i c_{2i}(E^c) \in H^{4i}(X, \mathbb{Z}).$$

Then $p(E)=1+\Sigma p_i(E)$ is the total **Pontrjagin class** of $E$.

(a) $f^* p(E)=p(f^* E)$ for $f: Y \to X$ in $\mathscr{C}$ which follows from the same properties for Chern classes.

(b) $p(L)=1$ for $L \in L(X)$ because $p_1(L)=-c_2(L^c)=0$.
Similarly, it is always true that the Pontrjagin classes vanish for dimensional reasons.

(c) $p(F \oplus Q) \neq p(F) \cdot p(Q)$ in general because $p(E)$ ignores the odd dimensional Chern classes $c_{2i+1}(E^c)$.
In fact $2c_{2i+1}(E^c)=0$ and therefore $p(F \oplus G)=p(F) \cdot p(G)$ modulo 2-torsion.

(d) $2c_{2i+1}(E^c)=0$.

***Proof.*** Since the coordinate transformations of $E^c$ are real, $E^c$ is equal to the bundle $\overline{E}^c$ obtained by taking complex conjugates. For complex line bundles $\overline{L}=L^*$, and so by the argument of Lemma 5, $c(\overline{E})=c(E^*)$ for all complex vector bundles $E$.
Thus $c_{2i+1}(E^c)=c_{2i+1}(\overline{E}^c)=-c_{2i+1}(E^c)$ and $2c_{2i+1}(E^c)=0$.
Recall the principal bundle of a compact Lie group and without the proofs we introduce some results of the universal principal $G$-bundle and characteristic cohomology classes. □

**Theorem 7**

Let $G$ be a compact Lie group.

(1) There exists a principal $G$-bundle $PG \to BG$ such that $PG$ is a contractible space.

(2) If $P \to X$ is any principal $G$-bundle, then there is a map

$f: X \to BG$ such that $P = f^*(PG)$.

(3) Two maps $f_1, f_2: X \to BG$ induce equivalent bundles if and only if $f_1$ and $f_2$ are homotopic.

**Definition 6**

Let $G$ be a compact Lie group.

A **principal $G$-bundle** over a space $X$ is a fiber bundle $P \to X$ with fiber $G$ whose coordinate transformation group is $G$ by a left action.

There is a canonical free right action of $G$ on $P$ and the orbit space $P/G = X$. We summarize the cohomology groups of some useful classifying spaces.

**Theorem 8**

(1) $H^*(BO(1), \mathbb{Z}_2) = H^*(\mathbb{RP}^\infty, \mathbb{Z}_2) = \mathbb{Z}[z]$, $z \in H^1(\mathbb{RP}^\infty, \mathbb{Z}_2)$.

(2) $H^*(BU(1), \mathbb{Z}) = H^*(\mathbb{CP}^\infty, \mathbb{Z}) = \mathbb{Z}[x]$, $x \in H^2(\mathbb{CP}^\infty, \mathbb{Z})$.

(3) $H^*(B(O(1) \times \cdots \times O(1)), \mathbb{Z}_2) = \mathbb{Z}_2[z_1, \cdots, z_n]$, $\deg z_i = 1$, $i = 1, \cdots, n$.

(4) $H^*(B(U(1) \times \cdots \times U(1)), \mathbb{Z}) = \mathbb{Z}[x_1, \cdots, x_n]$, $\deg x_i = 2$, $i = 1, \cdots, n$.

(5) $H^*(BU(n), \mathbb{Z}) = \mathbb{Z}[c_1, \cdots, c_n]$, $c_i =$ the $i$-th Chern class.

(6) $H^*(BSO(2m), \mathbb{R}) = \mathbb{R}[p_1, \cdots, p_{m-1}, e]$, $p_i =$ the $i$-th Pontrjagin class, $e =$ the Euler class.

(7) $H^*(BSO(2m+1), \mathbb{R}) = \mathbb{R}[p_1, \cdots, p_m]$.

(8) $H^*(BO(n), \mathbb{Z}_2) = \mathbb{Z}_2[w_1, \cdots, w_n]$, $w_i =$ the $i$-th Stiefel-Whitney class.

# Appendix B: Low-Dimensional Manifold

## §1. One-Manifold

An **interval** is a connected subset of $\mathbb{R}$ which is not a point. It may be finite or infinite; closed, open or half-open. Any interval is diffeomorphic either to [0, 1], (0, 1] or (0, 1).

**Theorem 1**

Any smooth, connected manifold of dimension one is diffeomorphic either to the circle $S^1$ or to some interval of $\mathbb{R}$.

Therefore there are only four distinct connected one-manifolds.

## §2. Two-Manifold

The classification of two-dimensional topological manifolds was known a hundred years ago.

**Theorem 1**

If $\Sigma$ is a closed two-manifold, then

(i) $\Sigma$ is homeomorphic to $S^2$, $T^2$ or a connected sum

$$\#_g T^2 = \underbrace{T^2 \# \cdots \# T^2}_{g}$$

if $\Sigma$ is orientable.

(ii) $\Sigma$ is homeomorphic to $\mathbb{RP}^2$, or a connected sum

$$\#_h \mathbb{RP}^2 = \underbrace{\mathbb{RP}^2 \# \cdots \# \mathbb{RP}^2}_{h}$$

if $\Sigma$ is non-orientable.

**Example 2**

Suppose $S \subset \mathbb{CP}^2$ is a smooth complex curve of degree $d$, given in $\mathbb{CP}^2$ as the locus of zeros of a homogeneous polynomial $f(z_0, z_1, z_2)$ of degree $d$. The restriction of the tangent bundle $T\mathbb{CP}^2$ to S has a decomposition

$$T\mathbb{CP}^2|_S = TS \oplus N_S,$$

where $N_S$ is the normal bundle of $S$ in $\mathbb{CP}^2$. Then the first Chern classes on $[S]$ are

$$c_1(T\mathbb{CP}^2)[S] = c_1(TS)[S] + c_1(N_S)[S]$$

and $3d = (2-2g) + d^2$, where $g$ is the genus of $S$.
We have the genus formula:

$$\mathrm{g}(S) = \frac{1}{2}(d-1)(d-2).$$

Thus from the classification above Theorem 1, $S$ is the Riemann surface of genus

$$\mathrm{g}(S) = \frac{1}{2}(d-1)(d-2).$$

**Theorem 3**

(1) Fundamental groups:

$$\pi_1(\underset{g}{\#} T^2) \cong \frac{< a_1 > * < b_1 > * \cdots * < a_g > * < b_g >}{< \prod_{i=1}^{g} [a_i b_i] >},$$

$$\pi_1(\underset{h}{\#} \mathbb{RP}^2) = \frac{< a_1 > * \cdots * < a_h >}{< a_1^2 \cdots a_n^2 >}$$

(2) Cohomology groups:

$$H^k(\underset{g}{\#} T^2, \mathbb{Z}) = \begin{cases} \mathbb{Z}, & k = 0, 2 \\ \mathbb{Z}^{2g}, & k = 1 \\ 0, & \text{other} \end{cases}$$

$$H^k(\#_h \mathbb{RP}^2, \mathbb{Z})=\begin{cases}\mathbb{Z}, & k=0\\ \mathbb{Z}^{h-1}, & k=1\\ \mathbb{Z}_2, & k=2\\ 0, & \text{other}\end{cases}$$

$$H^k(\#_h \mathbb{RP}^2, \mathbb{Z}_2)=\begin{cases}\mathbb{Z}_2, & k=0,\ 2\\ \mathbb{Z}_2^h, & k=1\\ 0, & \text{other}\end{cases}$$

## §3. Three-Manifold

In 1957 Papakyriakopoulos proved the three basic theorems of three-manifold topology: Dehn's Theorem, the Loop Theorem, and the Sphere Theorem. In dimensions 1, 2, 3, all topological manifolds are uniquely smoothable. Let $M$ be a three-manifold with boundary.

**Theorem 1 Dehn**

If $r\subset \partial M$ is a Jordan curve which is contractable in $M$, then $r$ bounds an embedded disk in $M$.

**Theorem 2 Loop**

In the Dehn's Theorem 1 the embedded curve $r$ is a loop in $\partial M$.

From Theorem 1 and Theorem 2, if the kernel of $\pi_1(\partial M)\to\pi_1(M)$ is nontrivial, then there is an embedded disk in $M$ whose boundary lies in $\partial M$ which represents a nontrivial element of this kernel.

**Theorem 3 Sphere**

If $M$ is orientable and $\pi_2(M)\neq 0$, then there is an embedded $S^2$ in $M$ which is not contractable.

The equivarient version of the theorems leads to a proof of the **Smith Conjecture.**

**Theorem 4**

If $\mathbb{Z}_n$ acts on $S^3$ with 1-dimensional fixed point set $F$, then $F$ is an unknotted knot.

In May 2006, a committee of nine mathematicians voted to award Perelman a Fields Medal for his work on the Poincaré conjecture. The Fields Medal is the highest award in mathematics; two to four medals are awarded every four years. However, Perelman declined to accept the prize. Nevertheless, on 22 August 2006, Perelman was publicly offered the medal at the International Congress of Mathematicians in Madrid "for his contributions to geometry and his revolutionary insights into the analytical and geometric structure of the Ricci flow." He did not attend the ceremony, and declined to accept the medal, making him the first person to decline this prestigious prize.

He had previously turned down a prestigious prize from the European Mathematical Society, allegedly saying that he felt the prize committee was unqualified to assess his work, even positively.

On 18 March 2010, Perelman was awarded a millennium Prize for solving the problem. On June 8, 2010, he did not attend a ceremony in his honor at the Institute Océanographique, Paris to accept his \$1 million prize. Perelman refused to accept the Millennium prize in July 2010.

**Theorem 5** **Perelman**

Every closed simply connected 3-manifold is homeomorphic to the 3-sphere $S^3$.

## §4. Four-Manifold

Throughout this section, $X$ denotes a closed simply-connected topological four-manifold. In paticular $X$ is orientable. Fix an orientation of $X$. There exists a natural symmetric bilinear form on the second homology group $H_2(X, \mathbb{Z})$, called the **intersection form** $\omega$ defined as follows. For $\alpha, \beta \in H_2(X, \mathbb{Z})$, let $PD(\alpha)$ be the Poincaré dual of $\alpha$ in $H^2(X, \mathbb{Z})$. Then

$$\begin{aligned}\omega(\alpha, \beta) &= <PD(\alpha), \beta> \\ &= \int_B PD(\alpha) \\ &= \int_M PD(\alpha) \cup PD(\beta) \\ &= A \cdot B\end{aligned}$$

where $<\ ,\ >$ is the Kronecker pairing between the cohomology group and the homology group. One known from the Poincaré duality theorem that the bilinear form $\omega$ is actually symmetric and unimodular (that is, any matrix representing $\omega$ has determinant $\pm 1$).

When $X$ is a smooth four-manifold, there is a well-known geometric way to define $\omega$. Represent $\alpha, \beta \in H_2(X, \mathbb{Z})$ by two smoothly embedded, oriented surfaces $A$, $B$ in $X$ such that $A$, $B$ intersect transversely.

Define $w(\alpha, \beta)$ to be **algebraic intersection number** of $A \cdot B$. Of course, there is still another equivalent definition of the intersection form using differential forms and wedge product on the cohomology $H^2(X, \mathbb{R})$. The intersection form is a basic topological invariant for closed four-manifolds.

In 1949, Whitehead showed that the homotopy type of a closed, simply-connected four-manifold is completely determined by the isomorphism class of the intersection form.

It is therefore important to know a little bit about unimodular symmetric bilinear forms on lattices (i.e., finitely generated free abelian groups, where $\omega$ vanishes on torsion classes).

Call two symmetric bilinear forms $\omega_1$ and $\omega_2$ on lattices $\wedge_1$ and $\wedge_2$ (respectively) **equivalent** or isomorphic if there is an isomorphism

$$\varphi : \wedge_1 \to \wedge_2 \text{ such that } \varphi^* \omega_2 = \omega_1.$$

A bilinear form $\omega$ on a lattice is called an **even type** if $\omega(\alpha, \alpha)$ is even for any $\alpha \in \wedge$. Otherwise, $\omega$ is called an **odd type**. There are two fundamental invariants of a symmetric bilinear form $\omega$ on a lattice: its **rank** (the dimension of $\wedge \otimes \mathbb{R}$) and its **signature** ($=$rank$-2q$, where $q$ is the maximal dimension of a subspace of $\wedge \otimes \mathbb{R}$ on which $\omega$ is negative definite). **Indefinite unimodular symmetric bilinear forms** are completely determined up to equivalence by their rank, signature, and type. The classification is as follows;

$$\omega \text{ is of odd type} : \omega \cong <1> \oplus m <-1> \ ; \ (n, \ m \neq 0)$$

$$\omega \text{ is of even type} : \omega \cong nE_\delta \oplus m \begin{bmatrix} 0 & 1 \\ 1 & 0 \end{bmatrix}; \quad (m \neq 0)$$

where $nA$ means the direct sum of the matrix $A$ $n$ times.

The classiffication of **definite unimodular symmetric bilinear forms** is much more difficult and one should not expect a full or useful answer here. A result of Eisenstein and Hermite states that there are finitely many isomorphism classes of positive definite unimodular forms of a given rank. However, the following statistics shows the difficulty in carrying out the classification; there is one even definite symmetric unimodular form of rank 8, 2 of rank 16, 24 of rank 24, and more than $10^{51}$ of rank 40.

To state the classification theorem, one needs the **Kirby-Siebenman invariant** $\alpha(X) \in \mathbb{Z}_2$ for a topological four-manifold $X$. It is defined as follows; $\alpha(X)=0$ if $X \times S^1$ is not smoothable, and $\alpha(X)=1$ if $X \times S^1$ is not smoothable.

**Theorem 1** **Freedman**

Simply-connected oriented closed four-manifolds are a 1-1 correspondence with pairs of the set

$$\{((\omega),\ \alpha) \mid \text{if } \omega \text{ is even, then the relation } \operatorname{sgn}(\omega)/8 \equiv \alpha \pmod 2 \text{ is imposed}\}$$

where $(\omega)$ is the isomorphism class of the form $\omega$, and $\operatorname{sgn}(\omega)$ is the signature of the form $\omega$.

Thus, every unimodular bilinear symmetric form is the intersection form of a simply-connected topological four-manifold. This manifold is unique when the form is even, and there are exactly two distinct manifolds when the form is odd.

**Theorem 2** **Donaldson**

The intersection form of a smooth simply connected four-manifold with positive definite form is equivalent to $(1)\oplus\cdots\oplus(1)$ over the integers.

**Example 3** All nonsingular algebraic surfaces $d$ of degree $d$ in $\mathbb{CP}^3$ are diffeomorphic and their basic algebraic invariants are given by;

(1) Signature $(V^d) = -\frac{1}{3}d^3 + \frac{4}{3}d$,

(2) Rank $H_2(V^d;\ \mathbb{Z}) = d^3 - 4d^2 + 6d - 2 =$ the 2nd Betti number $b_2(V^d)$,

$$H^1(V^d;\ \mathbb{Z}) = H^3(V^d;\ \mathbb{Z}) = 0$$

(3) The second Stiefel-Whitney class satisfies

$$\omega_2(T(V^d))(\in H^2(V^d,\ \mathbb{Z}_2)) = 0 \text{ iff } d = \text{even},$$

where $T(V^d)$ is the tangent bundle.

***Proof.*** The fundamental homology class of $V^d$ represents $d\hat{a} \in H_4(\mathbb{CP}^3;\ \mathbb{Z})$, where $\hat{a}$ is the canonical generator of $H_2(\mathbb{CP}^3;\ \mathbb{Z})$.

Let $a$ be the canonical generator of $H^2(\mathbb{CP}^3;\ \mathbb{Z})$.

Let $T$ be the tangent bundle to $\mathbb{CP}^3$ and $N$ an extension of the normal bundle of $V^d \subset \mathbb{CP}^3$ to all of $\mathbb{CP}^3$. The total Chern class of

$$c(T) = 1 + 4a + 6a^2 + 4a^3,$$

and

$$c(N) = 1 + da,$$

$$\text{so } c(-N) = 1 - da + d^2a^2 - d^3a^3.$$

By the sum formula

$$c(T-N) = 1 + (-d+4)a + (d^2 - 4d + 6)a^2 + (-d^3 + 4d^2 - 6d + 4)a^3$$

and

$$\begin{aligned} p_1(T-N) &= c_1^2 - 2c_2 \\ &= (-d+4)^2 - 2(d^2 - 4d + 6)a^2 \\ &= (-d^2 + 4)a^2. \end{aligned}$$

Since $T-N$ restricts to the tangent bundles of $V^d$, $TV^d$ we have

$$\begin{aligned} < p_1(T-N),\ [V^d] > &= p_1(T-N)|[V^d] \\ &= p_1(TV^d)[V^d] = p_1 V. \end{aligned}$$

But by Hirzebruch's formula;

$$\begin{aligned} \text{signature } (V^d) &= \frac{1}{3} p_1 V \\ &= < \frac{1}{3}(4 - d^2)a^2,\ d\widehat{a^2} > \\ &= \frac{-1}{3} d^3 + \frac{4}{3} d. \end{aligned}$$

Similarly,

$$\begin{aligned}\chi(TV^d)[V^d] &= <c_2(T-N),\ [V^d]> \\ &= <(d^2-4d+6)a^2,\ d\widehat{a^2}> \\ &= d^3-4d^2+6d.\end{aligned}$$

But the second Betti number satisfies $b_2 = x(\vee^{d})$ since

$$H_1(V^d;\ \mathbb{Z}) \cong H_3(V^d;\ \mathbb{Z}) \cong 0,$$

so $b_2 = d^3 - 4d^2 + 6d - 2$.

Finally $\omega_2$ is $(d^2-4d+6)a^2$ (mod 2)
and vanishes iff $d$ is even. □

# Appendix C: Category

## §1. Category

(1) A **category** $\mathscr{C}$ consists of

(a) a class ob$\mathscr{C}$ of **objects**,

(b) for each $A$, $B \in$ ob$\mathscr{C}$, a **set** Hom$(A, B)$ of **morphisms** from $A$ to $B$,

(c) for each $A$, $B$, $C \in$ ob$\mathscr{C}$, a composition

Hom$(B, C) \times$ Hom$(A, B) \to$ Hom$(A, C)$, $(g, f) \mapsto g \circ f$

such that

(i) for each $A \in$ ob$\mathscr{C}$ there is an identity $I_A \in$ Hom$(A, A)$, i.e., each $B \in$ ob$\mathscr{C}$, each $f \in$ Hom$(A, B)$ $f \circ I_A = f$, and each $f \in$ Hom$(B, A)$ $I_A \circ f = f$,

(ii) If $A$, $B$, $C$, $D \in$ ob$\mathscr{C}$, $f \in$ Hom$(A, B)$, $g \in$ Hom$(B, C)$, and $h \in$ Hom$(C, D)$,

then $h \circ (g \circ f) = (h \circ g) \circ f$.

(2) Examples of some categories:

(i) $\mathscr{Sets}$: objects : sets, morphisms : maps.

(ii) $\mathscr{Abgp}$: objects : abelian groups, morphisms : homomorphisms.

(iii) $\mathscr{Top}$: objects : topological spaces, morphisms : continuous maps.

(iv) similarly, the category *Gp* of groups, $R$-modules (for a fixed ring $R$), vector bundles on $X$ (for a fixed topological space $X$) etc.

(3) For given two categories $\mathscr{C}$, $\mathscr{D}$, a **covariant functor**

$$F : \mathscr{C} \to \mathscr{D}$$

is given by

(a) a map $F$: ob$\mathscr{C} \to$ ob$\mathscr{D}$

(b) each $A$, $B \in$ ob$\mathscr{C}$ a map

$F$ : Hom$(A, B) \to$ Hom$(F(A), F(B))$

such that

(i) for each $A \in \mathrm{ob}\mathscr{C}$, $F(I_A)=I_{F(A)}$,

(ii) for each $f$, $g$, $F(f \circ g)=F(f) \circ F(g)$ if $f \circ g$ is defined.

A **contravariant functor** has, instead of (*b*), (*b*′) each $A$, $B \in \mathrm{ob}\mathscr{C}$ a map

$$F: \mathrm{Hom}(A, B) \to \mathrm{Hom}(F(B), F(A)),$$

and (ii) is replaced by

(ii′) for each $f$, $g$, $F(f \circ g)=F(g) \circ F(f)$ if $f \circ g$ is defined.

(4) Examples of functor:

(a) The fundamental group functor

$$\pi_1 : \mathscr{Top} \to \mathscr{Gp},$$

and the homology group functor

$$H_n : \mathscr{Top} \to \mathscr{Abgp}$$

are covariant functors.

(b) The cohomology group functor

$$H^n : \mathscr{Top} \to \mathscr{Abgp}$$

is a contravariant functor.

(c) Let $X$ be a topological space, $\mathscr{U}$ an open covering of $X$ ordered by inclusion. A presheaf of abelian groups on $X$ is a contravariant functor $\mathscr{U} \to \mathscr{Abgp}$.

(5) Let $\mathscr{C}$ be a category of objects $A$, $B$, $\cdots$ and morphisms $\alpha$, $\beta$, $\cdots$.

$$\alpha : B \to A$$

is a **monomorphism** if the following condition is satisfied; given $\beta$

and $\gamma: C \to B$ such that $\alpha\beta = \alpha\gamma$, then $\beta = \gamma$.

Dually

$$\alpha: A \to B$$

is an **epimorphism** if, given $\beta$ and $\gamma: B \to A$ such that $\beta\alpha = \gamma\alpha$, $\beta = \gamma$.

$\alpha: A \to B$ is an **isomorphism** if there exists a morphism $\beta: B \to A$ such that $\alpha\beta = 1_B$ and $\beta\alpha = 1_A$. An isomorphism is both mono and epi, but the converse of this statement is in general false (e.g., in the category of topological spaces and continuous maps).

Let $\beta$: $B \to A$ and $\gamma$: $C \to A$ be monomorphisms. We write

$$\beta \leq \gamma$$

there exists a morphism $\alpha$: $B \to C$ such that $\gamma\alpha = \beta$ (and $\alpha$ is then unique). The relation $\leq$ is transitive on the class $M$ of all monomorphisms into $A$, $A$ being a fixed object of $\mathscr{C}$. If $\beta \leq \gamma$ and $\gamma \leq \beta$ then $\alpha$ is an isomorphism, and we write $\beta \sim \gamma$. $\sim$ is an equivalence relation on the class $M$. We choose a monomorphism out of each equivalence class, and we call the chosen monomorphisms the **subobjects** of $A$.

Thus a subobject of $A$ may be regarded as an object $B$ in $\mathscr{C}$ together with a monomorphism $B \to A$.

Dually, by considering the class of all eqimorphisms $A \to B$, where $A$ is fixed and $B$ variable, we define **quotient objects** of $A$: a quotient object of $A$ is an object $B$ in $\mathscr{C}$ together with an eqimorphism $A \to B$.

## §2. Additive Category

For each pair of objects $A$, $B$ in $\mathscr{C}$ let Hom($A$, $B$) denote the set of all morphisms $A \to B$. The category $\mathscr{C}$ is **additive** if the following axioms are satisfied:

(1) **Addition of morphisms** : Each Hom($A$, $B$) has an abelian group structure (written additively) and the composition of morphisms is bilinear, i.e.,

$$(\alpha+\beta)\gamma=\alpha\gamma+\beta\gamma \text{ and } \gamma(\alpha+\beta)=\gamma\alpha+\gamma\beta$$

whenever those compositions make sense.

(2) **Existence of direct sum** : Let $A$, $B \in \mathscr{C}$ then there are an object $C \in \mathscr{C}$ and morphisms $\alpha: A \to C$, $\beta: B \to C$ with the following property: given any morphisms $\alpha': A \to C'$, $\beta': B \to C'$ then there is a unique morphism $\gamma: C \to C'$ such that $\gamma\alpha = \alpha'$ and, $\gamma\beta = \beta'$.

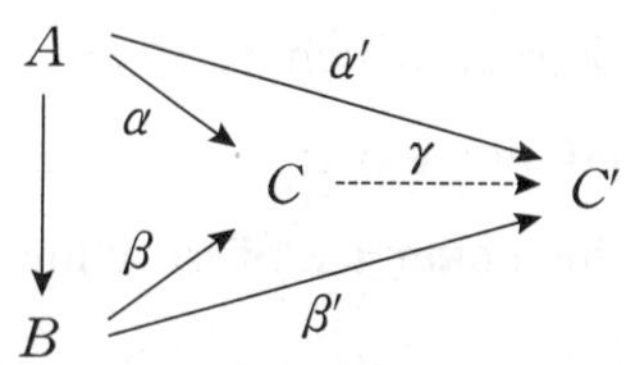


(3) **Existence of zero** : There is an object $A$ in $\mathscr{C}$ such that Hom($A, A$) consists of a single element, i.e., $1_A=0$.
The $A$ is called a **zero** of $\mathscr{C}$.

***Remark***

(1) The object $C$ of (2) together with the morphisms $\alpha$, $\beta$ is a **direct sum** of $A$ and $B$.
(2) One shows that any two direct sums of $A$ and $B$ are canonically isomorphic, and that (2) is in fact self-dual, i.e., it is locally equivalent to its dual, which formulates the existence of the direct product.
(3) Any two zeros of $\mathscr{C}$ are in fact canonically isomorphic.

Let $\alpha: A \to B$ be a morphism in additive category $\mathscr{C}$, and let

$$\mu: K \to A$$

be a monomorphism. $\mu$ is a **generalized kernel** of $\alpha$ if the following condition is satisfied:

$$\gamma : C \to A \text{ is such that } \alpha\gamma = 0 \text{ iff there is a morphism } \delta : C \to K \text{ such that } \gamma = \mu\delta.$$

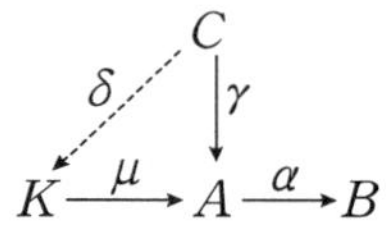


If $\alpha$ has a generalized kernel then the generalized kernels form an equivalence class in the class of all monomorphisms into $A$, hence there is just one generalized kernel which is a subobject of $A$, and this one is the **kernel** of $\alpha$ (It need not exist).

Dually we define the **cokernel** of $\alpha$ which (if it exist) is a quotient object of $B$; and then we define the **image** of $\alpha$ to be the kernel of cokernel of $\alpha$, which is a subobject of $B$ and dually the **coimage** of $\alpha$ is the cokernel of the kernel of $\alpha$, which is a quotient object of $A$. If $\alpha$ has an image and a coimage then $\alpha$ can be factorized canonically into

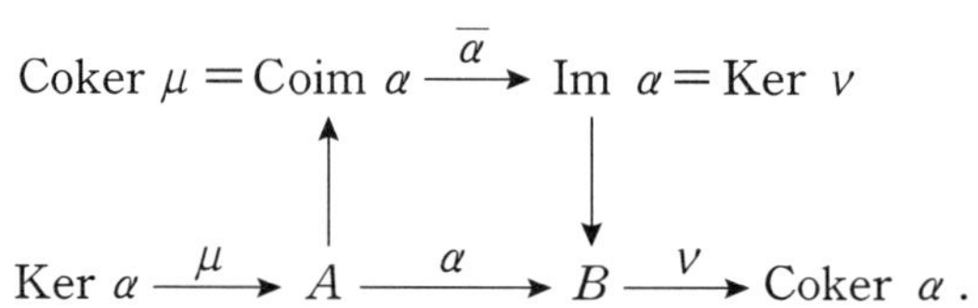

## §3. Abelian Category

An additive category $\mathscr{C}$ is **abelian** if

(1) every morphism in $\mathscr{C}$ has a kernel and a cokernel,

(2) if $\alpha$ is a morphism in $\mathscr{C}$ then

$$\overline{\alpha} : \text{Coim } \alpha \to \text{Im } \alpha$$

is an isomorphism.

As a consequence of (2), a morphism which is both epi and mono is an isomorphism (for if $\alpha$ is both epi and mono then $\alpha = \overline{\alpha}$).

**Examples of abelian categories**:

(a) The category of abelian groups and homomorphisms.

(b) The category of $R$-modules and $R$-homomorphisms, where $R$ is a fixed ring with a unit element.

(c) The category of presheaves of abelian groups on a fixed space $X$.

(d) The category of sheaves of $\vartheta$-modules on $X$, where $\vartheta$ is a fixed sheaf of rings on $X$.

There are many categories which are additive but not abelian, e.g., the category of topological abelian groups and continuous homomorphisms. Here (2) fails, for a continuous isomorphism need not have a continuous inverse.

# REFERENCES


(1) M. F. Atiyah and F. Hirzebruch, The Riemann-Roch theorem for analytic embeddings, Topology. 1 (1962) pp. 151～166.

(2) G. E. Brendon, Sheaf Theory, MacGraw-Hill Co., 1967.

(3) H. Cartan and S. Eilenberg and N. Steenrod, Foundation of Algebraic Topology, Prinston Univ. Press, 1952.

(4) Y. S. Cho, Differential Topology on Manifolds, Daewoo Science Book 432, Arche 2002.

(5) Y. S. Cho, Topology, Kyungmoon press, 2010.

(6) Y. S. Cho, Algebraic Topology, Kyungmoon press, 2010.

(7) G. de Rham, Variétés Differentiables, Hermann and Cie, 1955.

(8) S. Donaldson, An application of gauge theory to the topology of 4-Manifolds, J. Diff. Geom. (1983) pp. 269～316.

(9) M. Freedman and F. Luo, Selected Applications of Geometry to Low-Dimensional Topology, U.L.S., Vol. 1, A.M.S., 1987.

(10) R. Godement, Topologie algébriqe et théorie des faisceaux, Hermann, 1964.

(11) P. Griffiths and J. Harris, Principles of Algebraic Geometry, Wiley-Interscience Publication, John Wiley and Sons, 1978.

(12) A. Grothendieck, Sur quelques points d'algebre homologique, Tohoku Math. Journal, IX 1957, pp. 119～221.

(13) R. Gunning and H. Rossi, Analytic Functions of several Complex Variables, Prentice-Hall, Inc., Englewood cliffs, N.J., 1965.

(14) F. Hirzebruch, Topological methods in algebraic geometry, Springer, 1966.

(15) J. Milnor, Topology from the differentiable Viewpoint, University Press of Virginia, 1976.

(16) K. Stein, Introduction to the theory of several complex variables, Lecture notes Cambridge, 1961.
(17) R. G. Swann, Theory of sheaves, Univ. of Chicago Press.
(18) B. R. Tennison, Sheaf Theory, Cambridge Univ. Press, 1975.
(20) R. O. Wells, Jr., Differential Analysis on Complex Manifolds, Springer, 1980.

# INDEX

## 【D】



## 【E】

**【F】**



**【G】**



**【H】**

## 【O】



## 【P】



## 【Q】



## 【R】



## 【S】

【T】



【U】



【V】



【W】



【Z】

## Sheaf Theory



KM KYUNG MOON Publishers
121-818, 174 Wowsanro Mapo-gu Seoul, Korea
Tel: (82) 2 332 2004
Fax: (82) 2 336 5193

Printed in Korea

Price: ₩15,000